MILLENNIAL METROPOLIS

The text offers a critical perspective on complex and consequential aspects of growth and change in London, viewed through the lens of multiscalar space, and brought to life through exemplary case studies. It demonstrates how capital, culture and governance have combined to reproduce London, within a frame of relational geographies and historical relayering.

Emphasis is placed on the sequences of political change, capital intensification, industrial restructuring and cultural infusions which have transformed space in London since the 1980s. Tom Hutton contributes to the rich discourse on London's experiences of urbanization, by producing a fresh perspective on its development saliency. *Millennial Metropolis* includes a systematic review and synthesis of research literatures on globalizing cities, with reference to the reproduction of space at the metropolitan, district and neighbourhood scales. Hutton offers a nuanced treatment of geographical scale, observed in the blending of global/transnational processes with the fine-grained imprint of governance processes and social relations. These processes are manifested in sites of innovation, spectacle and social conviviality, but also produce experiences of displacement and inequality. The author presents a spatial model of metropolitan development by exploring how growth and change in twenty-first-century London is expressed internally as an enlarged zonal structure extending beyond the traditional territories of central and inner London. Serious threats to London are discussed—from the isolating implications of Brexit, the impact of the Covid-19 pandemic and the dire threat of ecological crises and deteriorating public health associated with climate change.

This will be an invaluable text for postgraduate students, established scholars and upper level undergraduates, across diverse disciplines and fields including geography, sociology, governance studies and planning and urban studies.

Tom Hutton is Professor of Urban Studies and City Planning in the Centre for Human Settlements, School of Community and Regional Planning, University of British Columbia, Canada. His research interests include new industry formation in the inner city, the role of service industries in urban transformation within the Asia-Pacific and the complex imprints of culture and social groups within the districts of the city.

MILLENNIAL METROPOLIS

Space, Place and Territory in the Remaking of London

Tom Hutton

LONDON AND NEW YORK

First published 2022
by Routledge
2 Park Square, Milton Park, Abingdon, Oxon OX14 4RN

and by Routledge
605 Third Avenue, New York, NY 10158

Routledge is an imprint of the Taylor & Francis Group, an informa business

© 2022 Tom Hutton

The right of Tom Hutton to be identified as author of this work has been asserted by him in accordance with sections 77 and 78 of the Copyright, Designs and Patents Act 1988.

All rights reserved. No part of this book may be reprinted or reproduced or utilised in any form or by any electronic, mechanical, or other means, now known or hereafter invented, including photocopying and recording, or in any information storage or retrieval system, without permission in writing from the publishers.

Trademark notice: Product or corporate names may be trademarks or registered trademarks, and are used only for identification and explanation without intent to infringe.

British Library Cataloguing-in-Publication Data
A catalogue record for this book is available from the British Library

Library of Congress Cataloging-in-Publication Data
A catalog record has been requested for this book

ISBN: 978-1-138-23248-8 (hbk)
ISBN: 978-1-138-23250-1 (pbk)
ISBN: 978-1-315-31249-1 (ebk)

DOI: 10.4324/9781315312491

Typeset in Bembo
by KnowledgeWorks Global Ltd.

For Lesley and Siena, as always; and to the memory of my mother,

Anne Margaret Burford Hutton, b. Islington London

CONTENTS

List of figures	*ix*
Preface	*xii*
Acknowledgements	*xiv*

1 Urban change and the reproduction of space: London
as field of study — 1

2 The governance of space in London: From political
economy to neoliberalism — 18

3 Money metropolis: Financialisation, spaces of capital
and the property machine — 40

4 Global capital and urban megaprojects: The appropriation
of place-identity and social memory — 63

5 Culture and the remaking of place and territory: Space,
local regeneration and the instrumental use of culture — 92

6 The convivial city: Spaces of spectacle, encounter
and experience — 119

7 Spaces of innovation: The 'Tech City' trope and place
branding in London — 142

viii Contents

8 New gentrifiers and emblematic territories:
Hypergentrification, displacement in situ and territorial
dislocation   169

9 Inscriptions of postcolonialism and social history:
Class signifiers, persistent localism and vernacular
spaces in London   195

10 Brexit London: The liberal city in a neoliberal state   218

Appendix   *235*
Bibliography   *244*
Index   *259*

LIST OF FIGURES AND TABLE

Figures

2.1	The county of London: Factories with over 100 workers in 1898	24
2.2	The county of London: Factories with over 100 workers in 1955	24
2.3	Clusters of East End tailoring, 1888	25
2.4	Clusters of East End tailoring, 1955	26
2.5	Industrial areas and product specialisation within the inner north-east London industrial district	27
2.6	The Greater London Council area and London boroughs	31
3.1	BNP-Paribas financial complex, Marylebone	51
3.2	Landscapes of fund management firms, Mayfair	51
3.3	Reconstructed landscapes and the built environment of the City of London	54
3.4	The new face and imagery of the City: A view from the south bank of the Thames	55
3.5	The corporate landscape of a global financial cluster: Canary Wharf	56
3.6	The Canary Wharf International financial complex	57
3.7	Street-level view of corporate office towers, Canary Wharf	58
3.8	The zonal structure of twenty-first-century metropolitan London	60
4.1	Central St Martins and adjacent consumption landscape, King's Cross yards	72
4.2	Megaproject clusters in East London Thameside	74
4.3	The Greenwich Peninsula project (view from the Isle of Dogs)	75
4.4	View of the City Island project site	76
4.5	London's maritime history invoked in place-remaking in Docklands	77

x List of Figures and Table

4.6	Renzo Piano's 'Shard', Southwark	78
4.7	The template of megaprojects in west-central London Thameside	80
4.8	View of the Nine Elms-Vauxhall site	81
4.9	Project(ing) market appeal: Branding of project design and exclusivity	82
4.10	View of the (former) Battersea Power Station, Wandsworth	85
4.11	View of the (former) Battersea Power Station chimneys	86
4.12	Chelsea Barracks street-level hoarding, Lower Sloane Street	89
4.13	Public display of curated industrial artefacts, Battersea project site	91
5.1	The Royal College of Music, South Kensington	99
5.2	The University of the Arts London, Pimlico	100
5.3	'Design takes London': Event signage at the Victoria & Albert Museum	104
5.4	Cultural line map of Camden: Lineaments of an institutional archipelago	106
5.5	Landscapes of specialised production and consumption: Clerkenwell c. 2005	108
5.6	'Apulia' Pugliese restaurant, Cowcross Street: Clerkenwell	110
5.7	Bermondsey Street, London Borough of Southwark: Structural features	112
5.8	The White Cube gallery: Bermondsey Street	114
5.9	Aestheticised building exteriors within heritage sites: Bermondsey Street	115
5.10	Artisanal consumption within the aesthetic corridor of Bermondsey Street	116
6.1	Portobello Saturday Market, Notting Hill	126
6.2	The Bermondsey Antique Market, Southwark	127
6.3	Cultural experience and convivial consumption at the Victoria & Albert Museum	129
6.4	Football mad, football crazy	133
6.5	Situating the Tottenham Hotspur Stadium, showing geospatial coordinates	134
6.6	Residuals of the sporting past: The SSE arena, Brent	135
6.7	Display of NFL football scarves, New Wembley Stadium concourse	135
6.8	The Den, Bermondsey, London SE16	136
6.9	Food stall, Spitalfields district	138
6.10	The Camden Lock Market	138
7.1	Clusters of specialised health and medical institutions in Central London	148
7.2	The Dyson School of Design Engineering, Imperial College London	150
7.3	Display of innovation in specialised wood applications, Victoria & Albert Museum	153

7.4	Hoxton and the Shoreditch Triangle, London Borough of Hackney circa 2004	157
7.5	Convivial spaces of the Shoreditch Triangle	158
7.6	Leonard Street, Shoreditch, leading to Great Eastern Street	158
7.7	The 'Silicon Roundabout': Intersection of Hackney and St Luke's, Islington	160
7.8	The 'business after business' scene in the innovation economy: Goose Island Brewpub, Shoreditch High Street	161
7.9	Signage proclaiming the advent of 'Tech City', Shoreditch	165
8.1	Waterloo Terrace, Barnsbury, London Borough of Islington	174
8.2	The aesthetic consumption-scape of gentrification: Upper Street, Islington	175
8.3	Landscape of housing estates, Arnold Circus, Tower Hamlets	177
8.4	Artisanal production, retailing and consumption, Calvert Street, Hackney	177
8.5	Highgate Village, London N6	178
8.6	Consumption landscapes in Lavender Hill, South Battersea, Wandsworth	181
8.7	Student housing for Middlesex University, Wembley, London Borough of Brent	185
8.8	Britannia Village, West Silvertown, Victoria Dock: Community site planning map	186
8.9	Britannia Village: Housing form and heritage landscape	187
8.10	Recreational amenity sites proximate to Britannia Village	188
8.11	High-density residential units, Victoria Dock, London Borough of Newham	189
9.1	Cosmopolitan landscapes of consumption: Peckham High Street	200
9.2	Manze Eel and Pie Shop, Peckham	201
9.3	'Everyday globalization': foodscapes of Queensway W2	204
9.4	Bandstand, Arnold Circus, London Borough of Tower Hamlets	207
9.5	Exmouth Market: Clerkenwell	209
9.6	Gazzano's Italian Deli and Cafe, Farringdon Road	211
9.7	The Clerkenwell Workshops	212
9.8	Culture, upgrading in place and the aesthetic of consumption: Cowcross Street	214

Table

| 2.1 | Production regimes and representative industries for London's inner city | 28 |

PREFACE

In *Millennial Metropolis* I offer a critical perspective on complex (and consequential) aspects of growth and change in London, viewed through the lens of multiscalar space.

An axiom of contemporary urban scholarship postulates that each city, regardless of size, location and development pathway, is worthy of study. This stance has stimulated a constructive opening up of the urban studies field to encompass cities situated within the Global South, 'left behind' cities peripheral to era-shaping circuits of capital and culture, and 'ordinary cities' variously defined. But London continues to occupy an important place within the diverse fields and research agendas of urban studies, owing to its instructive experiences of change over extended history as well as the active present.

The value of London as case study is situated in its recurrent demonstration of the socioeconomic and cultural consequences of industrial restructuring and place-remaking. These have included the City of London and the Port of London as defining features of Britain's global city trajectory in the nineteenth century; the space-shaping impacts of the London Underground from 1863; and legacy effects of the specialized light industry districts within inner north-east London, redeployed for the arts and culture, innovation industries, consumption and housing.

The development of Canary Wharf as a new international financial centre imposed upon the Isle of Dogs by Margaret Thatcher's Conservative Government pioneered an era of neoliberalism, and shaped a postindustrial trajectory for the British capital. A corollary 'territorial financialisation' experience was initiated by the privatization of public housing, manifested in a comprehensive gentrification of London's neighbourhoods, and producing a social structure increasingly favouring both established and emergent elites.

London's global city vocation is structured by high-margin specializations, notably finance, intermediate services, and the property sector. The British capital's development pathway is also co-produced by the complex residuals of empire and contemporary postcolonialism; by networks of social actors; and by the stimulus of the arts, knowledge creation and scientific innovation.

These storylines serve as critical precursors to contemporary experiences of growth and change. While restructuring in late twentieth-century London was shaped by the political mobilization (and increasing privatization) of space, the last twenty years have seen a substantial widening and deepening of capital relayering. Intersections of cultural values, technological innovation and higher education have produced diverse space-typologies across the metropolis. These comprise the subject matter of *Millennial Metropolis*, informed by extensive field research, a program of informant interviews and conversations with colleagues, and sustained engagement with the literature.

The growth of London's cultural economy beyond established sites and institutions in Camden, Westminster and South Kensington has brought into play new places of creative experimentation throughout the metropolis. 'First generation' cultural quarters colonized by artists in the 1980s, exemplified by Hoxton and the larger Shoreditch area, have transitioned to territories of digital innovation, demonstrating the power of upgrading and succession, and play a part in local regeneration strategies. Relatedly a lively economy of convivial consumption has emerged in London, involving new sites and social actors, but also presenting aspects of exclusion and marginalization.

I describe the territorial impacts of mega-projects in London arrayed along the Thames: a critical development corridor for London in this century as it has been since the city's initial settlement two thousand years ago. Transnational capital, and the appropriation of place-identity associated with the corporate enlistment of star architects, designers and marketing concerns, has transformed space progressively eastward from Canary Wharf, but also touches down in the western reaches of the Thames, including Battersea and Chelsea. Megaprojects in London have attracted global investors as well as 'new gentrifiers' who represent a significant intensification of the upgrading narrative.

The cumulative effects of these globalizing processes include an enhanced power projection for London within international circuits of cities. Growth and change in twenty-first-century London is manifested internally as an enlarged zonal structure extending beyond the traditional territories of central and inner London, and which I present as a spatial model of metropolitan development.

But there are serious threats to London's global city status, to its cosmopolitan identity, and its social vitality and public welfare. These include the isolating implications of Brexit—a self-inflicted wound upon Britain's body politic, its society and economy; the impact of the Covid-19 pandemic, which has produced massive financial damage as well as deep human costs; and the dire threat of ecological crises and deteriorating public health associated with climate change.

Tom Hutton
University of British Columbia
Vancouver

ACKNOWLEDGEMENTS

I want to acknowledge the expert creative assistance of Eric Leinberger, of the UBC Geography Department, who prepared the maps and graphics for this book, and offered insightful advice on design, content and presentation issues.

I'm indebted to the generosity of colleagues prepared to share their deep knowledge of London with me, including Peter Hall, Graeme Evans, Jo Foord, Andrew Harris, Andy Thornley, Andy Pratt, Max Nathan, Chris Hamnett, Tim Butler, and Susan Fainstein.

At UBC I benefitted from the shared insights of colleagues on contemporary urbanism in general, and London's distinctive place in the complex agendas of urban studies more specifically, notably David Ley, Trevor Barnes, Jamie Peck, Elvin Wyly and Mark Vessey.

I acknowledge Professor John Darwin, Nuffield College, and Penguin Books, for the passage that opens Chapter 3, from his *Unfinished Empire: The Global Expansion of Britain* (London: Penguin 2012).

Ordnance Survey (all rights reserved) for Figure 3.5. London Borough of Camden Creative Services/Arts and Tourism Team for Figure 5.4.

I acknowledge the Sports Director, Lillywhites Plc, for Figure 6.5.

Finally, I want to thank Andrew Mould, Senior Editor and Publisher, and Egle Zigaite, Senior Editorial Assistant and Production Manager, at Routledge | Taylor & Francis for their support, constructive feedback and (by no means least) patience over the term of this (and other) projects.

1

URBAN CHANGE AND THE REPRODUCTION OF SPACE

London as field of study

Prologue: Situating London within global urbanisation circuits

London was the first metropolis of the modern era and the first world city. London's development history has been shaped by political power, government agencies and private consortia associated with mercantile trade and the imperial project, national hegemony in intermediate banking and financial institutions, and a distinctive form of industrialisation from the nineteenth century onwards. The British capital's development has also been influenced by particularly rich milieux effects, comprising deep social capital, cultural diversity and network formation, shaped by migration as well as practices of established cohorts and actors.

These vocations underpinned national primacy, and London's ascendancy within circuits of power and influence among European capitals and trading cities, including over history are Antwerp, Amsterdam, Paris, Copenhagen and Hamburg. Following contraction both in the economy and population base over the 1970s, associated with Britain's overall economic malaise, and benchmarked by declining productivity, unfavourable terms of trade and more particularly the collapse of the Port of London and basic manufacturing industries, London has emerged in the twenty-first century as a global exemplar of regeneration.

Over extended history London's spatiality has been produced by factors which have included advantages of national primacy in the concentration of institutions of political control within a highly centralised state, and by industrialisation and world city formation in the nineteenth century, associated with the expansion of international trade and the management of empire. The construction of 13 mainline train stations over the last century and a half, and the development of the Underground from 1863 onward, greatly facilitated the physical growth of

DOI: 10.4324/9781315312491-1

2 Urban change and the reproduction of space

London well beyond the London County Council area, with new housing estates penetrating deeper within the metropolitan fringe and the Home Counties.

London's growth trajectory following deep industrial decline and overall population loss during the 1970s was influenced by a 'new international division of labour' (Fröbel, Heinrichs and Kreye 1980) within which apex-level global cities are privileged in the concentration of higher-order services industries, and with industrial labour shifting to the growth economies of East Asia. Forty years on from this prognosis London has largely retained its competitive advantage in intermediate services within an increasingly multipolar world urban order. London's economy also has strengthened in other key sectors underpinning contemporary metropolitan economies, exemplified by higher education and the knowledge sector, culture and creative industries and the innovation economy.

London's ascendancy as global city has been underpinned by concentrations of finance, corporate control and specialised services and labour in the City and Canary Wharf, and is expressed in new design values and idioms within reconstructed landscapes across the metropolis. London, along with New York, has been a principal beneficiary of the deregulation of financial markets, although the dividends have been disproportionately captured by executives, shareholders and fund managers and traders.

At the national level, Doreen Massey (1984) described a distinctively asymmetrical spatial division of labour in Britain which powerfully favours London over provincial cities in the distribution of high-wage occupations in finance and intermediate services: a tendency which has if anything strengthened in this century. London has also been the principal beneficiary among British cities of membership within the European Union, and has more to lose than other cities from a hard Brexit which would deny or obstruct entry to the single market. Similarly, possible restrictions on the free movement of workers, students and immigrants ensuing from Brexit would be damaging, as each is key to London's social capital and unique cultural vibrancy.

Platforms of development in globalising cities

Twentieth-century cities within leading national economies were largely shaped by a specific sector or industrial ensemble which defined eras of growth and change, including first the manufacturing sectors of the 'industrial city', and then the postindustrial city characterised by expanding services employment. Over the final three decades of the twentieth century, the collapse of traditional industries produced urban economies and labour markets within which services (both intermediate and final demand) accounted for a preponderance of employment growth.

The reassertion of London's growth trajectory and power projection in an era of transnational urbanism, and the ascendancy of cities among the dynamic economies of East Asia and the Global South which underpins an increasingly multipolar urban world, have been achieved on a robust platform of multiple high-margin industries, skilled labour and institutions that facilitate knowledge creation and innovation.

Like New York, San Francisco and Shanghai among other world cities, London has in this century capitalised on the development potential of the most propulsive sectors situated within advanced economies. These include notably the multiple platforms of intermediate finance and related services; cultural industries; higher education and the knowledge sector; and innovation across specialised fields of applied science, medicine and information technologies.

There are of course important differences in the governance systems of each of the bastions of the global economy, situated within the Atlantic realm of mature economies, the growth economies of East Asia increasingly led by China, and the emergent economies of the Global South. What most have in common though are variants of twenty-first-century capitalism which privilege the interests of the market and associated agencies and actors. A corollary experience of globalisation is observed in urban housing markets which cater increasingly to highly remunerated, knowledge-intensive workers.

A very large consumption sector of restaurants, cafés, bars, theatres and musical performance venues complements London's visitor economy, and offers opportunity for individuals imbued with talent and creativity. The consumption economy caters to the intensive social interaction needs of London's high-wage financial, technology and knowledge sectors. These activities and their diverse sites have in the aggregate contributed to the production of a more complex space-economy in the metropolis.

Internal markers of change within London

London's development trajectory has entailed sustained investment and enterprise formation within the established clusters of the central city, but has increasingly brought into play a far more extensive array of districts, sites and communities across the metropolis. Space is increasingly mobilised for development not only by corporations but also by the multilevel state and its agencies, by public and private institutions and by high net-worth families and individuals. Importantly the experience of change includes the serial reproduction of place and territory in response to technological innovation, market trends and preferences, and shifts in returns to capital, notably in the property sector.

Over extended history and in the active present London has benefitted greatly from inflows of immigrants, traders and refugees, enriching the capital's cultural milieu and social diversity, and deepening knowledge of commercial possibilities and external market dynamics. London's economy and social structure are highly internationalised: a consequence of London's development history, and sustained by massive flows of inward investment, the extensive presence of multinational corporations and subsidiaries across key sectors, and a diverse pattern of immigration matched only by New York.

Large numbers of young workers and students collectively animate the spaces and place-imaginaries of the metropolis, including participation in cultural industries and institutions, and the visitor economy and higher education. Young people also contribute in important ways to public welfare (and to lively cities) through engagement with social media, community associations and volunteering sectors.

London encompasses districts and communities which present a vernacular built environment motif, notably in prestigious areas of the central area. Much of central and inner London's built environment has its origins in the Victorian era (1837–1900) and, as Anthony King has observed (1976), the cultural implications of empire have included the imprints of colonial architecture in the British capital. Indeed the nineteenth-century world city experience shaped a comprehensive aesthetic for London, expressed in architecture, urban design, art and literature, as recounted in Celina Fox's volume (1992) cataloguing a major exhibition in Essen on 'London – World City 1800–1840'.

London overall lacks the coherent domestic architectural template of cities such as Paris, Florence, Amsterdam and Vienna among other cases. But the legacy of Victorian architectural values and idioms, building types and heritage spaces within many districts has proven to be conducive to accommodating the arts, cultural industries and creative workers, and more recently the innovation economy, demonstrating the robust utility of London's built environment over cycles of change.

In the contemporary era of neoliberalism (and inter-city competition), London has consciously opened up both to global capital and to design innovation from diverse international sources, as well as from British firms and professionals, observed in distinctive projects and iconic buildings situated across the metropolis. This impressive record of redevelopment and experimentation in architecture and urban design has been achieved in turn through comprehensive capital relayering and associated upgrading processes, contributing to widespread dislocation and deep inequality, as well as wealth and opportunity for many endowed with high levels of specialised knowledge and skills.

London's development storyline: Continuity and contradiction

My purpose in this volume is to contribute to a deeper understanding of London's developmental saliency within contexts of globalisation, urban transnationalism and industrial restructuring. As prelude I seek to identify illustrative meanings (and contradictions) of London within urban studies which necessarily have a bearing in the first instance on positioning the metropolis as field of study.

London performs as the capital of a mature, established unitary state. But London's political and economic primacy is a source of friction within the UK's diverse polities and regions, and is widely seen as an impediment to a more balanced pattern of regional development. London's size, prosperity and power projection within Britain were viewed as aggravating features in the Brexit vote of June 2016 and in the continuing political alienation of electorates in smaller and more peripheral British regions and towns, and was further underscored in the regional, community and class voting patterns in the 12 December 2019 British election.

Second, London was the largest industrial city within a nation that was the earliest to industrialise, but which also experienced the extensive collapse of both traditional industry and labour and the Port of London during the second half of the twentieth century, creating deep hardship and deprivation. Employment contraction undermined the residential tenure of industrial labour throughout much of the inner city, enabling extensive gentrification in traditional working class communities, as representatives of the new middle class exercised agency in processes of revalorisation and displacement. London's extensive former industrial districts and riverside docklands have been repurposed for large capital projects, high-value housing, the arts and cultural industries and for the innovation economy. Manufacturing now accounts for rather less than 5 per cent of London's employment, a proportion roughly equal to that of Hong Kong.

Third, London was the control centre for the world's most extensive empire for two centuries, an experience replete with subjugation of colonial populations and exploitation of resources, but has emerged as a critical site of postcolonial contestation and inequality as well as reconciliation in the postwar era. Immigration from former colonies contributes significantly to the shaping of a uniquely cosmopolitan urbanism, and to the productive diversity of London's social ecology. There is broad public acceptance of the many reciprocal benefits of multiculturalism in the British capital. But minority groups and individuals still face discrimination in daily life in contemporary London.

Finally, over the nineteenth and early twentieth centuries London was at the forefront of radical thinking and the formation of progressive movements, exemplified by the polemics of Karl Marx, the rise of Fabian socialism and imaginative thinkers such as William Morris and the growth of the Parliamentary Labour Party. The commitment of elected representatives in the London County Council (1888–1965) and Greater London Council (1965–1986) eras to public housing and to political economy values in the management of the metropolitan economy is a matter of record (see for example, Hall et al 1973; Thornley 1992; Travers 2003). But the last four decades have seen a steady rise in inequality in London measured in terms of incomes, public welfare and measures of health, as documented by such scholars as Chris Hamnett (2003, 2016) and Danny Dorling (2013). Indeed London represents among other things a pioneering site of neoliberalism within the global political landscape, replete with variants shaped by both national and local governments since the 1980s.

Key factors in the growing inequality in the metropolis include the exaltation of the market over social goods and community values as the leitmotif of Margaret Thatcher's Conservative government, rollbacks to local planning powers as a means of facilitating high-value development and the privatisation of much of London's council (public) housing stock.

6 Urban change and the reproduction of space

Successive governments, including both Conservative and New Labour, have tinkered with the neoliberal project but have largely accepted the basic order favouring marketisation over public goods. As Susan Fainstein observed in her influential treatise on *The Just City* (2003), London ranks well below the best European examples of progressive urbanism, the latter including notably Amsterdam, measured by key indices of effective representation, access to housing, social equality and inclusion.

Millennial Metropolis: An outline of purpose and objectives

My approach is not to insist that London uniquely stands for a particular moment as urban archetype; but rather that it offers an especially rich and instructive field for investigating the developmental implications of cities increasingly open to capital and to cultural penetration, and to relational geographies of change: through purposeful intent, or as an indirect outcome of broader changes in governance, regulation, policy and planning practice.

London's value as field of study is enhanced by its complex external relations, incorporating both competition and collaboration with cities and societies situated within a particularly extensive range of networks, including other apex global cities such as New York, Tokyo and Hong Kong. London's network interactions include economic, political and social relations with other British cities and communities, weighted by an extreme national primacy matched only by Seoul in the case of South Korea, and Bangkok, as capital of Thailand.

London benefits from a deep network of financial, business and cultural relations with other European capital cities and business centres, exemplified by Paris, Amsterdam, Berlin, Frankfurt and Milan. London's numerous universities and colleges sustain mutually enriching collaborations and exchanges with international universities, while the British capital's many cultural institutions similarly enjoy the benefits of reciprocal programs with counterparts in many countries.

London as former control centre of empire sustains political and social relations within the postcolonial realm and the Commonwealth, exemplified by institutional interaction with Mumbai, Cape Town, Lagos, Islamabad and Singapore among many other cities. Many of London's institutions, non-governmental organisations, political bodies and community groups are committed to processes of recognition and reconciliation with counterparts situated within former colonies. London also cooperates with cities committed to sustainable development programs though the C40 network (with Copenhagen and Vancouver as other examples), including roles for government actors and non-governmental organisations.

My purpose in Millennial Metropolis is to contribute to the rich discourse on London's experiences of urbanisation, with a view to producing a fresh perspective on London's development saliency. The book's content is historically informed, but emphasises the remarkable processes and imprints of change since

Margaret Thatcher's political revolution and radical economic program in the 1980s. Objectives of the book are as follows:

To track continuity and change in London's developmental trajectory in the late modern era as keys to an understanding of the transformation of space, place and landscape;

To retheorise the relational processes of development, growth and change in London in the reproduction of space and territory; and

To produce a critical assessment of London's positionality within key discourses of urbanisation and urbanism, and its value in the current retheorisation of the transnational metropolis, through a study of change viewed through the thematic lens of capital, culture and space.

Throughout the discussion governance is assigned a critical role in the (re) production of space in London, as agency of power exercised through executive policy, fiscal and monetary capacity, regulation, and, increasingly, as the institutional embodiment of the development state. London also stands as an instructive example of multilevel governance, within which policies and programs promulgated by central, metropolitan and borough governments, and special purpose agencies, individually and jointly shape development experiences.

Spatial registers of growth and change in London

Interdependencies between processes of growth and change and the reproduction of space continue to represent basing points of urbanisation and urbanism, and constitute an important stream of urban scholarship. Allen Scott's treatise (1988) on *Metropolis: From Divisions of Labor to Urban Form* represents an influential example, incorporating a depiction of the robustness of the industrial district over time shaped by production specialisation and labour formation. Robert Lewis's case study (2008) of industrialisation in Chicago discloses both the larger governance context and internal systems investment underpinning the factory networks of America's heartland metropolis.

The collapse of Fordist manufacturing and associated industries, labour and communities within western societies, and the coincident rise of advanced services industries and employment, generated both comprehensive restructuring of space in the city, and a rich critical literature. The cumulative impact of restructuring in the city over the last quarter of the twentieth century included not only a forcible reorganisation of labour and work, but also a reordering of space in the metropolis, with attendant privileging of a postindustrial, knowledge-based class popularised by Daniel Bell. Doreen Massey and Richard Meegan (1980, 1982), Bennet Harrison (1992) and David Ley (1996) number among influential contributors to this important discourse, and to a deeper understanding of the transformational impacts of restructuring on the social order in the spaces of the city. Chris Hamnett (2003), Tim Butler and Loretta Lees (Butler and Lees 2006)

8 Urban change and the reproduction of space

have shaped an important discourse on the deep social implications of labour force transformations in London, and more particularly to the forceful and complex experiences of dislocation at the district and community levels.

Trajectories of urban change: Points of departure

There are important differences in the ways co-dependencies of urbanisation (and urbanism) play out in the twenty-first-century metropolis. The emergence of the 'multiplex city' as described by Ash Amin and Stephen Graham (1997), shaped not by a single process of industrial restructuring but by a range of innovations, has produced transformational change across larger territorial spaces in cities and urban regions. Thus the fields of growth and change in the city-region include assuredly the central city and inner city—critical terrains for restructuring in the last century—but also the older inner suburbs and outer suburban areas, as acknowledged in a burgeoning scholarly literature on 'global suburbanisms', in which challenges to established theory and planning models which privilege the central city are elucidated by scholars such as Roger Keil and Elvin Wyly.

Second, along with these larger expanses of space in play within the metropolis are storylines of territorial change including the proliferation of cultural quarters, creative consumption areas and spaces of convivial interaction. The complex spatiality of the contemporary city is shaped by megaprojects, sites of technological innovation, public markets, designated ecological zones and sites of spectacle and experience. Each offers the potential for producing more convivial forms of urbanism, and a more comprehensive animation of urban space. But each also presents an insistent revalorisation and upgrading of urban space, place and territory, producing new experiences of exclusion and marginality as well as opportunity.

Third, these diverse typologies of space and territory are widely associated with successful cities in an era of globalisation and, further, represent essential elements of competitive advantage in an era of neoliberalism and competition. These archetypes can be traced across urban systems and cities within 'advanced', 'transitional' and 'developing' societies, as problematic as these terms may be, consistent with the influence of marketisation and pervasive policy mimicry of alleged 'best practice' (Peck and Theodore 2015), but are especially evident in the so-called 'apex cities' within circuits of capital, culture and labour.

New York, Chicago, San Francisco, London, Milan, Barcelona and Berlin come readily to mind as cases demonstrating both the instrumental and symbolic uses of space in restructuring processes under evolutionary capitalism, as articulated by Stefan Krätke in his critical essay on 'cities in contemporary capitalism' (2011). But research discloses similar tendencies in globalising cities such as Shanghai, Mumbai, Johannesburg and Rio de Janeiro (see for example, van den Berg 2015). And overall 'culture' represents not a fringe element of the metropolitan economy but rather a dynamic mélange of multinational firms, specialised production sectors, institutions, architectural idioms, public spectacle and consumption, shaped by a range of multi-level institutions and agencies (Grodach and Silver 2013; Hutton 2016).

Space, place and territory as framing concepts

London's redevelopment trajectory since the 1980s has been shaped by shifts in governance and policy values, and global city formation comprising cultural institutions and consumption industries and labour as well as finance and corporate power. London is an apex business centre and basing point of the global economy, and has been a principal beneficiary of Britain's membership in the European Union, along with other British regions and cities that have enjoyed the economic, social and cultural benefits of association in the world's largest and most successful union.

As postindustrialism shaped late-twentieth-century London, including pervasive social and spatial change, and the far-reaching reproduction of both wealth and deprivation among different occupational groups, financialisation in London extends well beyond the institutional banking world to encompass the structures and operating systems of the larger economy. The primacy of finance and privileges of wealth accumulation increasingly permeates communities, households and the lives of families, individuals and diverse social groups.

Accordingly scholarship on London's storyline over the past half-century of transformation includes critiques of change (and continuity) across complex fields, including governance and institutions, the structure and operating systems of the economy, the reformation of labour markets and occupations, ethnicity, social class, gentrification, cultural identity and postcolonialism and architectural values and the built environment.

Each of these fundamental domains of growth and change in London can be observed through the (multiple) lens of space, or, more categorically, *space*, *place* and *territory*. I would go further and propose that a deep investigation of the critical interdependency of process and space is quite central to a broader understanding of the nature of the metropolis, its development trajectory, and the conflicts and dislocations that are part and parcel of the twenty-first-century global city. I propose that 'space' is most usefully framed not only as the field of change, in essence the plane geometry upon which urbanisation takes place; but also

- As a salient arena of governance, planning and policy experimentation;
- As a resource for redevelopment, capital relayering, architectural innovation and landscape reformation;
- As a factor in shaping social and economic change, producing inequality and dislocation as well as opportunity;
- As a staging ground for human experience, and as repository of social memory; and, recurrently,
- as site of cultural experimentation and social engagement in globalising cities.

10 Urban change and the reproduction of space

These fields serve to underscore the conceptual centrality of multi-scalar space to an understanding of urban change and more particularly global city dynamics in the twenty-first century. Relatedly, power relations (configured by the state, society and more particularly capital and market values) reproduce space and systems of privilege and inequality.

Problematising space in the city: Place and territory in urban change

Space as articulated by influential scholars, such as David Harvey and Edward Soja, comprises a keystone of critical urban studies in the current era of neo-liberalism. Both for Harvey and Soja, 'space' is not simply the locus of change, but also a framing site of capital relayering and deep social dislocation in actual places, exemplified by Harvey's incisive work on Paris (2003), and Soja's influential discourse on Los Angeles (2000). For my London case study a spatial construct is problematised by reference to allied concepts of 'place' and 'territory': critical staging grounds for investigating conflicts and complements in the human use of space, and in conflicts between community and capital in the globalising metropolis.

Place represents a crucial concept of urban studies, directly relevant as lens and focal point of study for sociology, geography, anthropology, planning and policy studies. 'Place' is deployed in these literatures and discourses to denote sites and spaces imbued with human experience and meaning in the city, and is socially and culturally constructed, deconstructed and contested over time, by: the state, communities, social actors and increasingly market players.

There are divergent (and contradictory) meanings of place as shaped by the state, society and market players, as K. C. Ho (1994), and Brenda Yeoh and Lily Kong (1994; 1995), have demonstrated in the instructive case of Singapore. The early motivation for heritage building restoration in Singapore on the part of the state was to preserve historic areas and buildings as place-differentiating assets in tourist promotion, while community groups, professionals and academics emphasised the value of the built environment as marker of community history, social identity and struggle over colonial (and postcolonial) history. Similarly Jane M. Jacobs' storyline of place-making and conflicted experience and meaning in Spitalfields in *Edge of Empire* (1996) discloses the layered (and contested) quality of space-making and remaking in the heart of the colonial city.

Place is marked *in situ* by cultural and historical signifiers, and holds a place in social memory, shaped by the experiences and imaginations of groups and individuals. Contrasts abound in the security, permeability and reproduction of place in the city. In high-income, elite communities and neighbourhood, there is likely to be inherent robustness of place, as tenure is linked to income and access to wealth. But even these traditional sites of privilege are subject to penetration by transnational elites and high net-worth individuals and groups.

Rescaling territory in the metropolis: Reference points for London

Relatedly the idea of *territory* figures prominently in the discourses of governance and statecraft, including treatments of the classical exemplars of territory as expressions of political control as well as social and cultural organisation. There is a deep literature on the meanings and problematics of territory in the Westphalian (and more recently post-Westphalian) state, having to do with aspects of jurisdiction, control and conflict. But at a lower scalar register territory has deep resonance for cities and communities, and therefore for the agendas of urban studies.

Stuart Elden has contributed to this discourse in his influential monograph, *The Birth of Territory* (2013), in which he traces the origins and contemporary meanings of this critical concept of space and governance, and usefully references both early scholarship and more recent work. Elden asserts that in his view Jean Gottmann's *The Significance of Territory* (1973) is the best general treatment of territory as spatial organisation of the state. In Gottmann's *La Politique des Etats at leur Géographie* (2005) he asserts that 'one cannot conceive a state, a political institution, without its spatial definition, its territory' (quoted in Elden 2013: 6).

Gottmann's oeuvre includes deep studies of the Mediterranean world of trade and development (1990), as described in an essay by Luca Muscarà which links Gottmann's 'Atlantic transhumance' to his influential spatial theory (1998); but also incorporates his highly original treatment of 'megalopolis', the urbanised northeastern seaboard of the United States: arguably his unique contribution to an understanding of the larger territorial expression of urbanisation in the twentieth century. Elden also reference Robert Sack (1983), who 'effectively argues that territoriality is a social construct, forged through interaction and struggle, and thoroughly permeated with social relations', emphasising the role of culture in shaping understanding of the workings of capital in space and time (Elden 2013: 4–5).

Territory, ethnography and urban change at the local scale

The emergence of new urban spaces shaped by social groups, cultures and innovative work practices, infused with signifying features of multiple sites, forms the materièl for new and lively narratives of territory in the metropolis. The spirit of this resurgence of territory in urban change and in ethnographic scholarship is captured in influential monographs of evocative American case studies. Michael Indergaard's *Silicon Alley: The Rise and Fall of a New Media District* (2004), offers a penetrating account of New York's Mid-Manhattan economy in the early years of the twenty-first century, associated with fusions of 'digital technology, edgy creativity and fast money' (back cover), a phenomenon of 'paper wealth' which 'crashed' and led to the reassertion of New York's platform of Wall Street, real estate and the media.

As notions of place in the city are replete with differentiating aspects of location and siting, social history, scale and cultural signifiers, the concept of *territory*

12 Urban change and the reproduction of space

in the contemporary metropolis is shaped by a range of scalar reference points. First we can identify at the highest scalar register the zonal structure of the metropolis, incorporating in many cases the central city; Central Business District (CBD) Fringe and inner city; inner and outer suburbs; and exurbs. For the British capital there are the traditional demarcations of central, inner and outer London, which still retain some analytical value.

Territory in the metropolis is also shaped by strategic policy and programs, as designated in larger development programs and projects. In London, Alan Ainsworth (2010) has written about the importance of innovation in the remaking of Clerkenwell, a traditional territory of specialist artisanal production over history situated on the City Fringe, and which takes in multiple sites, places and neighbourhoods.

The classic postwar example of a comprehensive reconstruction of territory is the Canary Wharf project: an almost totalising remake of a deprived social landscape of the metropolis forcefully reproduced by capital and neoliberal governance. In this century a succession of megaprojects has transformed London's landscapes, retaining emblematic buildings and heritage aspects and gesturing to social history, but for the most part promoting a steep social upgrading.

Land use policies and plans for 'regeneration' (Imrie, Lees and Raco 2009) are pitched at different scales for London boroughs (and partnerships between boroughs), but often encompass larger territories which may incorporate more detailed site plans, exemplified in the complex territorial form of the Nine Elms-Vauxhall-Battersea megaproject. Then there are the extended designations for iterations of the Thames Gateway plans, which are inchoate although notionally aspirational across a territorial expanse downstream from London.

Political economy offers an effective entrée into important insights in normative study of territorial formation in the city, including critical studies of the human presence, practices, rituals and interactions that demarcate and define territory. Territory is a fungible concept, but is effectively configured by neighbourhoods and districts as key markers and units for analysis. The intersections of the state, governance and social actors within space constitutes a fertile research terrain. Then there are the many markers of territory in the city, associated with deprivation experienced by social collectives, as elucidated by Willmott and Young (1957) in the case of Bethnal Green; and with districts characterised by specialised production processes and labour: a feature of territorial formation over London's history.

Territory is deployed as a spatial frame for redevelopment schemes and projects across the metropolis. But there are multiple (and multi-layered) forms and expressions of territory which shape the spatiality of the city. To illustrate there are diocesan areas that form territories of congregants and adherents, including in London population of diverse faiths and religious observation. London's principal football clubs, including notably Chelsea, Arsenal and Tottenham, encompassed communities of fans and followers with a distinctive geography shaped by proximity to each team's stadium. And in the most recent period a vibrant youth culture of convivial consumption and assembly has shaped territorial forms of social behaviour in selected areas of London.

The centrality of place and territory to the London narrative of growth and change

What has made London distinctive among mature, globalising cities is the remarkable extent of space and territory available for higher-value redevelopment at the close of the twentieth century. This space includes not only the very extensive docklands area downstream from Tower Bridge, and the industrial lands situated in much of inner north-east London, inner north-west districts such as Kilburn, and parts of South London; but also working class communities and neighbourhoods rendered vulnerable to displacement through a combination of industrial disinvestment and associated employment contraction.

Emphasis in Millennial Metropolis is placed on the sequences of change in governance, capital relayering, industrial restructuring and social structure that have transformed space in London since the 1980s. At a localised register, there are numerous spaces and territories within London delineated in literary and artistic works and in cultural realms over history. These include the districts of London which constitute the social terrains described in the diaries of Evelyn and Pepys, and later those of Disraeli, Waugh and Spender; artists' neighbourhoods such as Chelsea and Pimlico; Virginia Woolf and others members of the Bloomsbury circle situated in Camden; and filmmakers and artists in Soho and Shepperton.

Space, place and territory also comprise important features of London's socioeconomic development, embodying interdependencies of occupation, class and residential tenure. In the nineteenth century the brutal conditions of working class life in London were expressed powerfully in the trenchant novels of Charles Dickens, notably in districts of the City Fringe and east end. Wilmot and Young's resonant study of Bethnal Green brought to life the working conditions and hardships ameliorated by the strength of community.

At a strategic scale space is a critical feature of London's development experience, acknowledged in influential theory of urban primacy and control. London was one of seven city-regions included in Peter Hall's landmark study of World Cities (1966), the others including Paris, the Randstadt, the Ruhr, Moscow, New York and Tokyo. These city-regions were included on the basis of national economic primacy, industrial specialisation and urban scale. Hall's discussion of London emphasised the rise of white-collar, office-based services concentrated in the City of London, just as the metropolis's specialised industrial production sectors and labour were about to experience decline and collapse, in the process subverting the balance of sectors and spaces in the capital's economy.

Saskia Sassen's widely cited *Global Cities* (2001 [1991]) study which privileged New York, London and Tokyo as cities at the peak of a new international order of finance and corporate control, situated within central business districts in

14 Urban change and the reproduction of space

each case. In this century scholars working within the institutional framework of the Global and World Cities (GaWC) project at Loughborough University have significantly broadened the taxonomies of cities and spaces associated with global city formation, bringing in social, cultural and institutional measures of saliency.

I acknowledge David Harvey's ideas (1982, 2001) concerning space in the 'geographies of capitalist accumulation', and capitalism's need for a 'spatial fix' for situating the production-consumption nexus crucial to generating surplus, as point of entry for understanding London's experience of growth and change. Harvey's seminal contributions to our understanding of urban change also includes his incisive account of the saliency of banking, finance and capital to the forceful clearances and redevelopment of the right bank of central Paris in the nineteenth century as precursor to the serial reproduction of space in the twentieth century (Harvey 2003; see also Sutcliffe 1973). In Harvey's view, 'capital' is not simply a factor of production but rather a forceful agency of change and inequality in the metropolis.

I also draw on scholarship which offers nuanced understanding of the 'species of space' in cities, notably Mike Crang and Nigel Thrift's edited volume *Thinking Space* (Routledge 2000) which incorporates depictions of 'metonymic space', where certain cities stand for particular moments in urban history. These metonyms include Peter Hall's description (1998) of Berlin in the Weimar period as the European 'capital of culture', an extended lineage of scholarship which established Chicago as hallmark of an eponymous school of social ecology from the 1920s, and more recently the idea of Los Angeles as exemplar of postmodern urbanism, as proposed by Edward Soja, Michael Dear and other scholars.

Two decades into the twenty-first century the rise of China has effectively brought in Shanghai, and the instrumental use of space and territory in its globalisation trajectory, as exemplar of contemporary urbanisation (Zhong 2011), while a more inclusive urban studies has brought to the fore metropolitan cities of the Global South, including Mumbai, Johannesburg, Sao Paulo and Kuala Lumpur. In consequence more nuanced treatments of space which recognise aspects of contingency as well as general processes are clearly indicated. And here London's complex postcolonial relations with the former outposts of empire encompass another important agenda for critical study.

Mark Crinson's edited volume on *Urban Memory: History and Amnesia in the Modern City* (Routledge 2005) offers evocative narratives of memory and experience in the formation of meaning associated with urban *place*, especially with reference to postindustrial legacies situated within the built environment which form the essential matrix for much of London's redevelopment over the past four decades. Broadly 'place' applies to space in the city which embodies meaning for social groups and individuals, associated with ethnicity and class, as well as work practices. Contemporary scholarship increasingly brings into the discourse associations of place associated with gender, identity and demographic divisions, exemplified in Linda McDowell's scholarship (McDowell 2012, 2015).

Place-making has become an imperative of capital, multinational corporations and the local state, capitalising on the (often layered, discordant or contested) meanings and imaginaries of places in the city. This reading of continuity and conflict in both the material and semiotic meanings of history has special resonance for London, as observed in the redevelopment and upgrading experienced in the capital's extensive postindustrial territories, and in the remaking of long-established public and wholesale markets, and train stations and their proximate terrains (Knox 2011).

In *Millennial Metropolis* I engage with traditional forms of local area study in the metropolis, as well as the new ethnography of place and territory situated at the intersection of industrial innovation, culture and place. The research lineage in this field includes Rebecca Solnit's critical study of turn of the twentieth-century San Francisco, subtitled as 'the crisis of American urbanism' (2000); Richard Lloyd's study of 'neo-Bohemian' creatives and allied social groups in the Wicker Park area of Chicago (2006), K. C. Ho and Hutton's comparison of creative workers in Singapore's Chinatown and Little India districts (2012); and Murray Mckenzie and Hutton's essay on conflictual aspects of culture-led regeneration in Vancouver's Victory Square district (2015). This line of inquiry seeks to align knowledge concerning the social history and place-identity of districts and communities, with the practices and values of new actors who value at some level the local habitus as crucible of cultural expression and industrial innovation.

Logics of organisation and structure of the book

In Millennial Metropolis I examine complex processes implicated in the reproduction of space and territory in London. Following this introduction I present two chapters which establish the critical frameworks of change in London. Chapter 2 outlines the parameters of the extraordinary shift in 'policy distance' from postwar commitments to governance and planning in the broader public interest, to the contemporary political values privileging the market in the reproduction of space and site in London.

Chapter 3 ('Money Metropolis') depicts the far-reaching impacts of capital within London, its permeating effects on social groups and communities through insistent relayering and upgrading, and the implications of capital deepening for the reproduction of sites and landscapes across the metropolis: essential prelude to the subsequent substantive chapters and case studies. I establish through references to key contributors to the financialisation discourse, the power of capital to reproduce space across ever-increasing expanses of the capital: a twenty-first-century model of development in London that decisively favours the interests of capital and social élites. I conclude Chapter 3 with presentation of a spatial model of metropolitan London that takes in the new territorial structure of London, shaped by capital, strategic industrial clusters and social elites.

Subsequent chapters explore impacts of redevelopment across London, framed within instructive narratives and case studies situated within different registers of place and territory.

16 Urban change and the reproduction of space

I start with 'Global capital and megaprojects' (Chapter 4), acknowledging London's pioneering status in the progression of large capital projects from the nineteenth century, to the remarkable extent of current sites and projects in London. The storyline encompasses the scale of capital sourcing, the selective appropriation of local histories, and erasure of struggles of social groups, producing new landscape imaginaries that contribute to territorial change in the metropolis. A parallel narrative includes local regeneration programs that aspire to enabling greater opportunity for public benefit ensuing from redevelopment in London.

In Chapter 4 I develop descriptions of two principal territories of redevelopment along the Thames riverside and corridor: first, the evolution of redevelopment sites eastward from the pioneering Canary Wharf global financial site in Docklands, to contemporary projects in the Greenwich Peninsula and Victoria Dock. A second territorial framing of major capital projects encompasses redevelopment sites in west-central London, including the massive Nine Elms-Vauxhall-Battersea project in Wandsworth, as well as smaller but instructive (and space-shaping) projects across the River in Chelsea.

The instrumental use of culture in London, both as centrepiece of regeneration programs, and as inducement to higher-margin redevelopment, forms the narrative for Chapter 5, 'Culture and the Remaking of Place and Territory'. The value of culture lies in large part in London's status as apex global site of cultural institutions, along with cities such as Paris, St Petersburg, Hong Kong, New York and Washington DC, and comprising museums and galleries, performance spaces and libraries and special collections of cultural assets situated within districts and clusters.

But there is a more complex storyline to acknowledge in terms of how culture interacts with (and co-produces) space in London. Culture is acknowledged as stimulus to gentrification, and device for triggering upgrading in place, observed in the absorption of culture in London's innovation economy: reproducing territorial form, structure and identity.

Chapter 6 offers commentary on 'The Convivial City', which follows a field of study that asserts the saliency of consumption, entertainment and social interaction in the city. In London conviviality is observed in (and experienced within) the capital's many retail, restaurant and performance spaces—accessible in some form to large numbers of London residents and visitors. But there are also forces for territorial rebranding and dislocation at work, observed in upscale consumption, notably, within the capital's uniquely extensive sports clubs and stadia that also anchor larger clusters of residential development and amenity: reflecting larger forces of the globalisation of sports and changes in affiliation.

Chapter 7 ('Spaces of innovation and technology') carries forward this storyline of succession and reterritorialisation in London, observed in the capital's innovation economy at different scales and locations. The geography of innovation in London includes the capital's centrepiece institutions engaged in applied science, medicine and digital communications, including major research-based universities such as Imperial College, teaching hospitals and many labs and consultancies.

More recently London's status as a leader in the innovation economy has been substantially enhanced, including investment by major US multinationals, notably Google and Facebook. But the geography of innovation in London is more diverse, and, further, is linked to complex processes of succession and upgrading. Here I describe the recasting of established areas of arts and culture in East London as sites of contemporary innovation, including the role of multilevel governance in technology based regeneration.

Earlier chapters of this volume describe the progression of the gentrification narrative in London, including the proliferation of new groups and cohorts in serial upgrading of communities and neighbourhoods. The storyline includes both the intensification of social upgrading in place, as well as the increasing reach of gentrification beyond established sites of privilege. Chapter 8 offers a discussion of 'new gentrifiers and emblematic territories', which incorporates the dislodging of traditional elites in London by high-net-worth transnationals in bastions of privilege in Westminster and Kensington, as well in more peripheral locations, suggesting in turn a reordering of social relations and ideas of hierarchy in the global metropolis.

Capital relayering, intensification and dislocation form the central narrative of *Millennial Metropolis*. But following the work of scholars engaged in research on the persistence of vernacular cultures and the practices of 'everyday globalization' in the metropolis, elucidated by Timothy Shortell (2014) in his comparative study of Paris and Brooklyn, I describe in the book's penultimate chapter imprints of postcolonialism and social history (Chapter 9). There are still in the British capital resonant spaces of diverse social character that accommodate change at a lower register, offering access and opportunities for intercultural engagement for residents and visitors. These include diverse communities of immigrant populations, precincts of affordable consumption and vestiges of London's traditional working class.

In my final chapter (Brexit London), I offer an essay which comprises elements of synthesis derived from the preceding chapters, as well as conjecture on wider implications of the London case study. Discussion includes a periodisation of capital relayering and spatial change at the strategic level in London; a rehearsal of the relevance of standard spatial models to the London case; and finally a proposal for scaling the most salient features of contemporary London at three levels: platforms, territories and sites. In a coda to Chapter 10 I address the importance of space and territory in London associated with the highly differentiated human experiences of plague and pandemic in 1665 and 2020.

Millennial Metropolis concludes with an appendix, describing methods and sources associated with the research and writing of the book.

2

THE GOVERNANCE OF SPACE IN LONDON

From political economy to neoliberalism

Introduction: Governance, the state and society in urban change

Urban theory, notably the Chicago School of Social Ecology, and industrial urbanism models which link divisions of labour to urban form, has emphasised the developmental saliency of the internal configuration of urban space shaped by local and regional factors. These space-shaping factors include market geographies, the distribution of economic activity and impacts on urban form shaped by clusters of firms.

The social geography of the city is shaped by community patterns of occupation, class and ethnicity, and by innovation in residential building type and architectural form. These internal divisions within the city reflect gradients of class, privilege and inequality, as well as industrial specialisation and related functional qualities, and are associated with the reproduction of urban landscapes and place-imaginaries. Space typologies in the city are recognised and formalised by the local state, occasionally amended and managed through a suite of regulatory, land use and development policies and programs.

For many cities, exogenous influences on the larger development trajectory and remaking of place constitute well-established factors; these include typically capital cities, major international gateways and trading complexes and international financial centres. London's global city repertoire encompasses each of these specialised typologies, and therefore represents a prime example of a city for which the 'external' comprise exigent factors in the reproduction of space.

The spatial impress of London's complex relationships with the states, societies and markets of successive multinational developmental and trading regimes, a sequence including the North Sea, the Mediterranean, the Atlantic realm and then progressively the globalised circuits of empire, is distinctive among world cities and in some respects unique. In the late modern era, shifts in British

DOI: 10.4324/9781315312491-2

government attitudes toward Europe are invoked as benchmarks of policy change, from the generally distancing postures that characterised the early post-war period, to membership in the European Economic Community (as of 1973), and more recently to formal notice of departure (31 January 2020).

In this chapter I present a discussion of key processes implicated in the reshaping of space, place and territory in London. These include notably changes in governance, consequential shifts in policy and planning for London and the far-reaching impress of financialisation as development trajectory: a theme I address in Chapter 3, and which together with socio-cultural factors represent essential conditions for the quite remarkable reproduction of space in this exemplary metropolis.

The storyline includes reference to the policy and planning experience of the 1970s: representing a critical baseline for understanding the dimensions of change in the governance for London in the neoliberal era. The narrative presents a measure of the extraordinary 'policy distance' travelled between from the state commitment to strategic planning, regulation and development control prior to 1980, and the comprehensive market orientation of the twenty-first century established by Margaret Thatcher's Conservative Government four decades ago.

I reference initiatives aspiring to more effectively coordinate policy and planning between central agencies and local government in the London South East Region in the 1970s. Efforts of national government to micro-manage the economy in an era of industrial crisis and increasing inter-regional socioeconomic inequality are illustrated by the operation of commercial office development control in London. I draw on exemplary shifts in central government policies for London as a reflection of changing values, priorities and contradictions as the central state has at times aspired to manage London's primacy within Britain or, more recently, to deploy London's global city status in the national interest.

I identify postindustrialism as a process which defines in large part the economic trajectory of advanced economies in the 1970s and 1980s, shaped by neoliberal governance and the diminished political influence of unionised manufacturing workers, and by a new international division of labour (Fröbel, Heinrichs and Kreye 1980) favouring industrial production among the growth economies of East Asia, with western cities specialising increasingly in intermediate services.

But deindustrialisation lingers in the psyche and social memory of communities, planners and urban studies scholars, as Susan Christopherson has observed (2003). Postindustrialism also persists in the territorial patterns of former districts which have performed as sites of serial upgrading and redevelopment, and also in the deployment of industrial metaphors, signifiers and branding by property concerns in London and other global cities.

Rethinking the state-city dialectic: Currents and cross-currents

While London's development has been shaped by external relations over history, critical scholarship has offered perspectives on changing relationships between cities, regions and the larger state which ramify for the British capital. In

20 The governance of space in london

State-Space, edited by Neil Brenner, Gordon MacLeod, Bob Jessop, and Martin Jones (2003), writers offer insights on the broad contours and problematic features of governance since the deep restructuring experiences of the 1970s. An influential lineage of scholarship dating from Henri Lefèbvre's classic study on the production of space (1974), and Edward Soja's exhortation for a critical spatial perspective in social theory (2000), has shaped what the editors describe as a necessary critique of the 'taken-for-granted linkages between state, territory and society' (Brenner et al 2003: 3).

In the view of Brenner et al, the collapse of North Atlantic Fordism from the 1970s led to a crisis of the postwar welfare state, Keynesian management and the integrity of national government policy realms: the 'state as "power container" appears to have been perforated … and thus the inherited model of territorially self-enclosed, state-defined societies and economies has become highly problematic' (Brenner et al 2003: 3). In consequence at least two broad tendencies can be discerned: a 'new localism' which implies the 'reassertion of the importance of the local in economic regeneration, political participation, and community-building'; and, second, a 'new regionalism' which implies a repositioning of metropolitan cities and regions as 'strategically more competitive' units within developmental and trading systems (Brenner et al 2003: 3).

London is just one of a number of metropolitan cities whose economic power and cultural influence enable a significant level of projection across global capital circuits and basing points of development. Cities also exercise agency through membership in numerous international associations and institutions, including those concerned with trade, culture, social development and sustainability values. And national governments and state agencies have heavily invested in flagship cities such as Paris, Milan, Seoul and Beijing as elements of programs for global repositioning, and for attracting capital and skilled labour.

But there are though complications to this narrative which must be acknowledged. Not least among them is the election of national leaders such as Boris Johnson and Donald Trump whose mandates are largely derived from non-metropolitan towns and regions, and whose political messaging portends a marginalisation of the liberal, internationalist voice associated with progressive urbanism. This matters as it represents a key dialogic of government intent, although Trump's defeat in the US election of 3 November 2020 opens up possibilities for a re-balancing of investment and labour market development in America, with President-elect Joe Biden promising progressive policies for postindustrial heartland regions and cities.

In Britain the Conservative manifesto for the election of 12 December 2019 endorsed as policy mantra reinvestment within 'left-behind' communities and regions of the North and Midlands, and in the West Country. This commitment will be stringently tested in the aftermath of the 2020 and 2021 coronavirus pandemic, which in addition to the catastrophic human cost is likely to compromise government finances for a decade or more.

But if this program of redirecting state investment from the metropolitan South East Region to the North and Midlands is realised, it would represent a strategic shift from the policy values over an extended sequence of Conservative and 'New Labour' governments, as expressed in a sustained record of massive public investment in London's infrastructure, institutions and built environment.

The production of space in London: Markets, society and the state

London's development chronology is shaped by the establishment of institutions and agencies associated with finance, trade and political control, producing the territories of the City of Westminster and the City of London, enabling competition for markets with European rivals, and the financing of European wars and eras of colonialisation from the seventeenth century.

In sequence the scope of trade, competition and conflict expanded from the North Sea and cities and states situated on its littoral, and marked by the succession of Dutch wars in the seventeenth century; to contact and engagement with the states and societies of the Mediterranean in the time of Philip II. Fernand Braudel notes visits by English ships to Leghorn as early as 1573, with increasing trade and British naval presence from the seventeenth century onward (Braudel 1966: 624); and then to the Atlantic era of colonisation and the stimulus to global trade initiated by exploration, notably Cook's navigational feats in the Pacific (Darwin 2012).

A political and administrative class in London was formed as an important social correlate of the externalisation of the capital's external trading vocation, concerned with the management and diplomatic arrangements of exploration, trade, finance and colonisation. Over extended periods of conflict with continental states and more especially the succession of wars with France, London performed as hosting place for diplomacy and statecraft. These efforts were rudimentary at first, but over time comprised more formal arrangements and the establishment of embassies, consular services and trade agencies: reshaping space in capital cities as well as socio-cultural relations between states and societies.

The management of land and space in London is complex, owing to the intermingling of local and external factors of change. Through extended historical periods, London's spatial structure has been shaped (or influenced) by mercantile trade, flows of capital, and the landscape ramifiers of the imperial project.

There are also spaces in London which to an extent are alienated from the control of London government and administration. These include the sites of central government, including Parliament as well as many ministerial and departmental agencies in Westminster and elsewhere; ecclesiastic sites and institutions; spaces occupied by the British military command; residences of the royal family; and the capital's many foreign embassies which are in effect part of each respective sovereign state's territory.

Immigration represents historically and in the modern era a process of socio-cultural relayering in London, in response to pressures of political and ecclesiastical oppression within source societies. The settlement of some 20,000 Huguenot refugees (c. 1670–1710) in the Spitalfields area marked an important facet of London's economic, socio-cultural and spatial development, followed by successive flows of immigrants and refugees from an increasingly diverse array of states and societies over the past three centuries.

Immigration from Commonwealth societies in the postwar era greatly enriched London's social diversity, and reshaped cultural values and practices at the community level. More recently, generous provisions of entry to Britain for citizens of the European Union as part of treaty obligations enabled many individuals and groups from the continent to work, study and experience life in Britain's cosmopolitan capital, adding to London's cultural cachet.

With Parliament's formal notification of Britain's departure from the European Union as of 31 January 2020, with an 11-month negotiation period to follow, London's many and mutually enriching networks of trade, travel, culture and epistemic communities with European partners may be constrained by the terms of association at the larger state level. Britain's departure from the EU may also constrain formal responses to recurrent cultural and political conflict, refugee flows, the security (or porosity) of borders and more particularly the Covid-19 pandemic—a human catastrophe which has cruelly exposed the frailties and follies of governance in many nations. These failures of governance certainly include Britain, which has ranked among the worst of European nations in addressing the Covid-19 pandemic, evidenced by high rates of infection and deaths among the British population, although a successful vaccination program has produced much better outcomes in 2021.

London's experience of growth and change in the modern era

Governance and multilevel agencies of the state represent essential features of the London storyline of growth and change, and in particular the evolution of the state-space nexus. The central narrative takes the form of a tradition of London government in managing and regulating space in the metropolis, with highly consequential change for the practices of land management over the last century. The London County Council's (1888–1965) three-quarters of a century span of governance encompassed the expansion of London's economy, population growth and the production of important new spaces: including sites of industrialisation, residential development and community formation and public institutions and their constituent domains.

Much of London's growth over this period entailed expansion of residential populations well beyond the LCC boundary, greatly potentiated by the development of rail capacity and main line stations. In response to the continued expansion of the capital region, a new planning model designed to contain London's spatial growth in the interwar and postwar years was introduced, incorporating development control measures, a program of new and expanding towns and the establishment of a Metropolitan Green Belt on London's periphery.

There are important features of continuity observed across the 130 years of metropolitan governance for London since the establishment of the London

County Council in 1888. These include the financing of transportation infrastructure across an expanding metropolis, the funding of education and social services and the management of the built environment. The profit potential of commercial development and high-value housing has represented motive forces in the reproduction of space in the metropolis.

In this regard, Richard Dennis (2008) has written about what he describes as 'Babylonian flats' in Victorian and Edwardian London, using as case study the experience of Queen Anne's Mansions, situated between Victoria Street and St James Park. As he recounts pressures exerted by developers on public agencies (in this case the Metropolitan Board of Works established 1855) for extra scale in terms of building envelope and elevation associated with demand from affluent clients is a near-constant for London in the modern era.

Victorian London encompassed a dramatic expansion of the metropolitan space-economy and critical industries. Some of these, notably in the finance and banking sectors, survive as vital platforms for London's twenty-first-century economy, while others have withered in the face of competition, obsolescence, product cycle effects and pressures of the London property market. But each has left its mark on London's landscape and social history.

The most prominent examples include the City of London and its specialised financial institutions and industries, and which I address in the following chapter on financialisation and the role of capital in landscape formation; allied with the London docks which have been comprehensively transformed through a mixture of adaptive reuse and massive capital relayering over the past four decades: the subject of Chapter 4 in this volume.

Industrial London: Spaces, territories and legacies

London's traditional industrial vocation, although subsumed in many narratives of London's development by high-margin banking, finance, corporate control and intermediate service industries, represented a significant element of the metropolitan economy from the 1840s to its collapse in the 1970s. Further, the distinctive spatiality, built environment and imaginaries of industrial London still resonate, in the physical form of landscapes, in the marketing of residential properties, and in the recent shift from cultural industries to digital innovation enterprise in East London.

To illustrate the territorial range of industry and firm density at the metropolitan scale (defined by the boundary of the London County Council), Figure 2.1 shows the distribution of factories with 100 workers or more in 1898, near the end of the Victorian era. Figure 2.1 discloses dense clustering of firms in central and inner London north of the Thames, and with significant representation in districts such as Southwark, Millwall and Camden Town.

Figure 2.2 reveals the pattern of factory location in London in 1955, with some dispersion from the Central London clusters evident but still in the aggregate demonstrating that manufacturing was important in mid-century London.

24 The governance of space in london

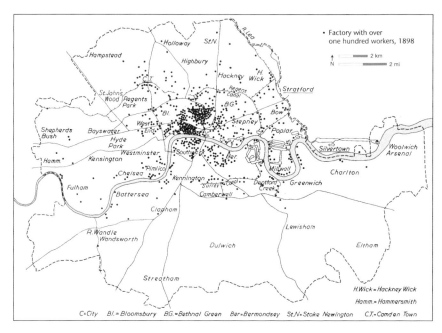

FIGURE 2.1 The county of London: Factories with over 100 workers in 1898
Source: Royal Commission on London traffic, Vol. V (1906) (from Martin 1964)

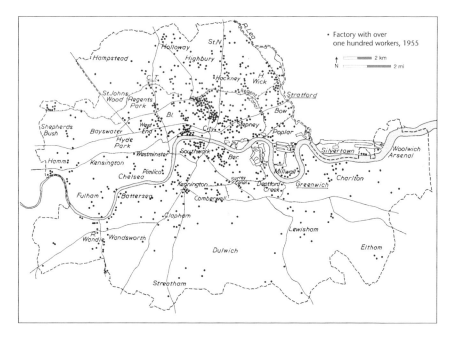

FIGURE 2.2 The county of London: Factories with over 100 workers in 1955
Source: Martin (1964)

The governance of space in london 25

At a finer level of industrial specialisation and spatial representation, Figures 2.3 and 2.4 show the changing cluster formation of tailoring in London's East End in 1888 and 1955, comprising an important sector and spatial division of labour within a globalising London over two periods. As for the broader representations of factories in Figures 2.1 and 2.2, the trend in tailoring business discloses some attrition in absolute numbers of enterprise, but with tailoring still clearly comprising an important industry in London in the mid-twentieth century.

The factories shown in Figures 2.1 and 2.2, and businesses situated within the dense congeries of tailors shown in Figures 2.3 and 2.4, have long disappeared from London's production landscape, supplanted by high-value products and services,

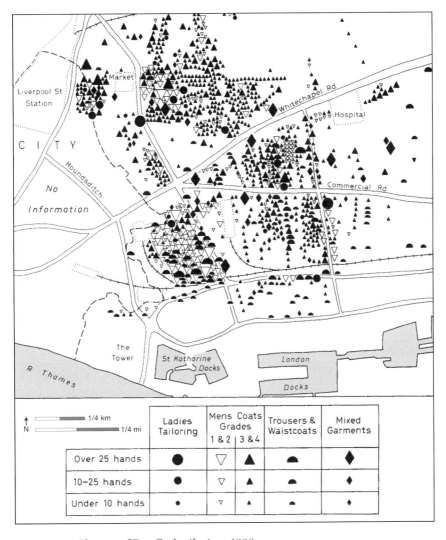

FIGURE 2.3 Clusters of East End tailoring, 1888

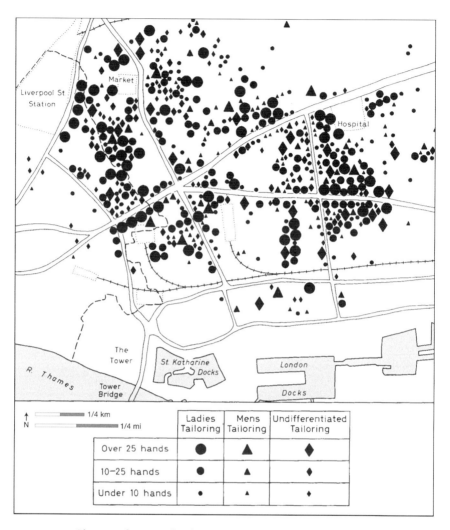

FIGURE 2.4 Clusters of East End tailoring, 1955

Source: Martin (1964)

and new labour formation. But the former spaces of specialised production retain deep saliency in twenty-first-century London. The complex patterning of industrial districts in East London and the City Fringe (Figure 2.5) comprise strategic sites of contemporary innovation, cluster formation and changing divisions of labour.

These districts have in many cases functioned as staging grounds for recurrent innovation sequences, new social groups and emergent divisions of labour. Defining processes from the 1980s have included transitions from artists' studios and galleries, to professional design firms, then to clusters of enterprise and labour in the 'innovation economy'; and in this century accompanied by high-value consumption and amenity.

The governance of space in london 27

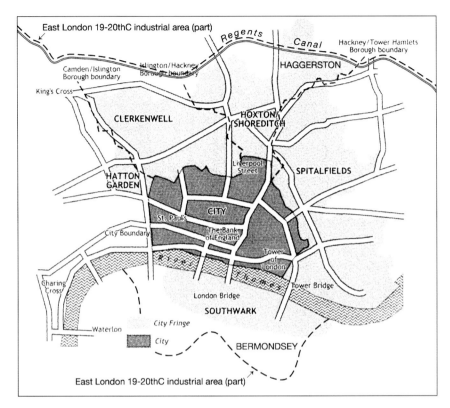

FIGURE 2.5 Industrial areas and product specialisation within the inner north-east London industrial district

Source: Martin (1964)

I depict the contours of production regimes in contemporary London, including representative industries within (residual) pre-Fordist, Fordist and importantly post-Fordist industries in Table 2.1, embodying the structural complexity of industry in the twenty-first-century global metropolis.

These industries are clustered in London's central area, the 'City Fringe' adjacent to the City of London, and the extended inner city including both the inner north-east sector and selected South London boroughs close to the Thames. These territories comprise the crucial staging grounds for innovation, restructuring and upgrading, which I describe in succeeding chapters of this volume.

Experiments in regional development and planning institutions

Britain's contribution to regional planning innovation at mid-twentieth century was exemplified by the introduction of the Metropolitan Green Belt and complementary programs for new and expanding towns for the larger London region,

28 The governance of space in london

TABLE 2.1 Production regimes and representative industries for London's inner city

I *Pre-Fordist industries*	*II* *Fordist industries*	*III* *Post-Fordist industries*	
		A *Intermediate* *service industries*	*B* *Cultural / New* *economy industries*
(1) Skilled artisans, artists, apprentices	(1) Operatives: skilled, semi-skilled labour, managers and supervisors	(1) Segmented labour: executives, managers, professionals, technical, sales and clerical	(1) Specialized neo-artisanal labour, design professionals, scientific and IT staff, artists, sales, managers
(2) Workshops, shops, residential space	(2) Factories and plants	(2) Office buildings	(2) Studios, workshops, live-works, work-lives, offices
(3) Artists Bespoke tailors Bookbinders Jewellers Milliners Model-builders Musical instrument makers Perfume and scent makers Precision instrument makers Silver plate engravers	(3) Food and beverage production • bakeries • breweries • food-processors Garment production (long-run, mass market) • factories and plants • sweatshops Printing and publishing • mass-market integrated Fordist production	(3) Corporate control: head and regional offices Intermediate banking and finance • merchant banking • fund managers • stock exchange • insurance Producer services • legal firms • accountants • marketing • management consultants Property • development companies • property managers • estate agents • research and market intelligence	(3) New media/multi-media Internet services and web-design Computer graphics and imaging Software design Digital arts Graphic design and arts Digital publishing and printing Film production and post production Video game production Music studios Galleries: curatorial services Specialized food and beverages • organic food • specialty bakeries • specialty coffee roasting • ethnic foods and beverages

Note:
(1) labour.
(2) infrastructure.
(3) representative industries.

Source: Hutton (2010)

The governance of space in london **29**

articulated in Peter Hall's magisterial two-volume set on *The Containment of Urban England* (PEP 1973): a planning model emulated with local variations throughout much of the urbanised world. 'Containment' occupies an important place in the lexicon of urban-regional planning as an effective approach to addressing problems of over-concentration, urban sprawl and dispersed development.

By the 1970s new institutions were required to advance regional planning in Britain, as the state policy agenda was increasingly shaped by problems of decline as well as aspects of growth in some sectors. Even the most prosperous British regions compared unfavourably with leading continental regions in terms of industrial investment and GDP measures.

This era was pithily described by Jamie Peck as encompassing 'a multisectoral downturn in UK manufacturing after 1968' (Peck 2013: 101). Peck cites Doreen Massey's trenchant analysis of manufacturing decline in the UK, derived from a careful analysis of 31 industrial sectors, as comprising at the national level three 'conceptually distinct' processes: 'productivity-enhancing intensification of existing production systems; investment-driven technical change, entailing a reorganisation of these systems; and the rationalisation of employment and capacity' (101).

A marked regional dimension to the national economic crisis included heavy impacts on cities in the North and Midlands, which specialised in traditional resource sectors (notably coal-mining) and manufacturing industries. Pressures on traditional industries included increasing international competition, disinvestment, product cycle effects and disastrously bad labour relations in important cases, notably mining and the auto sector.

London's economy vis-à-vis those of other urban-regional economies within the UK, while more diversified than those of the regions, and enjoying the advantages of higher shares of growth industries within the services sector, nonetheless shared in the economic malaise. The experience of industrial restructuring in London was unique in Britain, comprising several strategic-scale effects. In addition to losses of employment in London's extensive light industrial sector of specialised production noted earlier, the decline and collapse of the Port of London, for over a century the world's largest, was associated with the rise of containerised trade, with contractions in seaborne trade in the UK, and with infrastructural obsolescence. As Peter Hall observed, metropolitan London experienced a loss of 800,000 manufacturing jobs between 1961 and 1985 (Hall 1998: 889).

The accession of Britain to membership in the European Economic Community in 1973 was viewed as prospective stimulus to trade and industrial development amid national economic crisis, and encouraged thinking about complementary agencies at the regional scale. One such initiative concerned new machinery for promoting economic development through a more coordinated regional governance model. For the London and the South East Region, the government established the South East Economic Planning Council (SEEPC). This advisory body offered constructive advice to government but lacked key powers essential to promoting economic development on the model of leading continental states.

30 The governance of space in london

A second institutional innovation designed to improve regional planning and development took the form of the Standing Conference on London and South East Regional Planning (SCLSERP), representing the local authorities (comprising London boroughs and districts within the South East). The remit of SCLSERP included coordination of development across the region through institutional dialogue and collaboration.

There was though an emphasis on 'restraint' (rather than 'development') as policy dialogic associated with many of the districts within the South East Region, a mark of commitment to preserving the character of smaller communities within areas proximate to the Green Belt. A survey I conducted of district councils in the South East Region 1985–1986 disclosed in a majority of cases a local preference for 'restraint' (read as favouring stringent development control practices) even in sub-regional areas where the broader policy guidance encouraged higher levels of new investment and redevelopment (Hutton 1991).

Office development in London and Britain's regional problem

London was a principal site of manufacturing collapse within Britain over the 1970s. But London benefitted from growth in commercial office development, and the expansion of attendant professional, managerial and clerical employment—growth occupations within the ascendant services sector—to a greater extent than other British cities. London's capture of a preponderant share of office employment within Britain represented both a policy issue, as described in Doreen Massey's *The Regional Division of Labour* (1984), and political problem, as the socioeconomic effects of regional disequilibrium in office development between London and the rest of the UK became known: a condition which has persisted in the first two decades of the present century.

The central government's response to this aspect of regional divergence favouring London was twofold. The establishment of a Location of Office Bureau (LOB), initiated in 1963, was designed to publicise the opportunities for commercial office development in Britain's regions (Hall 1972). Second, a new regulatory system for office development was set up by the Department of the Environment, in the form of an Office Development Permit (ODP) system. Applicants for new offices in excess of 3,000 square feet were asked to demonstrate why they needed to locate in London, rather than in a provincial city (this was similar to an office permitting system in France designed ostensibly to deflect new development from Paris and the Ile-de-France to the provinces).

Although the threshold for exemption from review appeared very low, with 3,000 square feet about the size of a small estate agency or legal office, critics observed that the actual operation and oversight of the ODP system was insufficiently stringent to deflect enough new development to make a significant difference to the regional imbalance of employment growth. Further, the LOB was seen by scholars as ineffective in persuading developers of the merits of considering a location in provincial British cities (Daniels 1978).

Overall, the modest efforts of central government through LOB and the ODP system to control or at least manage office development in the London South East Region presented at best a light screening function, and made very little impact on the waxing economic dominance of the London region within Britain. Growth in London's office sector and intermediate service industries over the last three decades of the twentieth century comprised key underpinnings of the capital's dominant financial and business services sector, a trajectory continuing through to the present century, and situated both in new and established sites and spaces in the metropolis.

Tribulations of the Greater London development plan

The 1963 Greater London Government Act greatly expanded the old London County Council (LCC) area to incorporate (as Outer London) local authorities in the region with connections to the capital's labour market and employment centres, replacing the LCC with a new Greater London Council (GLC).

The establishment of the GLC (1965) was motivated by a central state recognition of the need for a genuinely metropolitan government encompassing London's territory (Figure 2.6). Accordingly, the GLC's territory extended well

FIGURE 2.6 The Greater London Council area and London boroughs

32 The governance of space in london

beyond the limits of the LCC, although not capturing all of London's contiguous settlement area.

There were a number of elements to the GLC's remit, including an extension of London government's traditional commitment to the provision of housing, public transportation and other infrastructure. The strategic policy project took the form of the Greater London Development Plan (GLDP), an exercise designed to offer guidance to managing the capital's economy and physical development at a time of deep industrial contraction, the growth of the commercial office sector in London and new housing estates and communities in outer London boroughs.

GLDP staff developed an office strategy encompassing modest growth in business centres such as Hammersmith and Croydon, mostly to serve local needs for business services. Office growth would continue in the City of London—a quasi-autonomous local government body which many critics on the left identified as an unrectified anomaly in the Greater London Act (1963).

Advice from academics, policy specialists and trade unions included concerns that an unrestrained office sector would increasingly dominate London's space-economy, generating inflation in land rents, deflecting capital from the renewal of the capital's manufacturing sector and undermining the sectoral balance of the economy increasingly in favour of services. For many critics, London's office sector operated as a vehicle for property speculation and (indirect) displacement of worthwhile and valuable but lower-margin industries. Centre Point at the eastern end of Oxford Street was designed explicitly as vehicle for capital appreciation, and remained vacant for many years—to many critics the epitome of the property based economy and its excesses.

Over its remit the GLDP project incorporated many of the values of progressive policy making in London at a strategic level. Knowledgeable staff were assigned to the project, and applied themselves diligently to the complexity of the policy development process, which included a significant research program, as well as outreach to academic experts, diverse communities and the 32 London boroughs (Massey 2007).

But the GLDP project must be considered an honourable fail. Lack of a coherent political direction for London's development future can be considered one issue, as was the problem of trying to bring the boroughs on board for the difficult choices required for the implementation of a development plan for the metropolis as a whole. The autonomy of the City of London represented a considerable impediment to the GLDP, with such a large and high-margin economic cluster outside the remit of the policy process. The two most consequential aspects of change—the collapse of traditional industry and the growth of commercial offices associated with London's global city role—could not be effectively managed by the Greater London Council and its limited policy capacity.

There is also the aspect of the extended time horizon over which the GLDP process was undertaken. As Francis Sheppard noted, the plan took four years to prepare; three years for the public inquiry phase; and four more years for the

Secretary of State to approve the plan in modified form (Sheppard 1998: 349): the very antithesis of the 'fast policy' model which characterises the neoliberal model of statecraft.

Globalisation, restructuring and the policy problematic

From the deep industrial restructuring of the 1970s, and the growth of the commercial office sector, to the GLDP's travails in the 1980s, London was transitioning from the 'world city' model introduced by Peter Hall (1966) in which national primacy was a benchmark of inclusion, toward the 'global city' nomenclature developed by Saskia Sassen (1991). The global cities prospectus privileged primacy in corporate control, finance and intermediate services as measures of status in international rankings.

The concomitant rise of the new middle class and the devalorisation of working-class communities shaped a destabilising gentrification experience throughout much of inner London. Along with this tendency communities of artists colonised postindustrial areas of the metropolis, injecting cultural vibrancy into many of London's neighbourhoods, while adding to pressures for social upgrading: a theme to be picked up in Chapters 5 and 6.

Changes in the warehousing and distribution of food products and other consumer goods were precursors to the repurposing of London's traditional wholesale markets to places of convivial consumption and social interaction in the city, exemplified by Covent Garden. And the first experiments in the adaptive reuse of former warehousing and dockside buildings along the River were undertaken, notably in Butler's Wharf in Southwark.

And in West London, Heathrow continues to experience high levels of growth in passenger throughput, driven both by increasing business travel associated with London's waxing global city status, and by international tourism, as well as a burgeoning air freight industry facilitating trade in high-value products. A growing complex of maintenance operations, offices and logistics operations grew up in the former Middlesex area proximate to Heathrow in Southern Hillingdon. But overall, the big drivers of London's economy comprised the specialised financial and diverse service ensembles clustered in Central London.

Remaking London in the neoliberal era: Margaret Thatcher and Michael Heseltine's agenda

The election of Margaret Thatcher's Conservative Government in 1979, at the end of a decade of national economic crisis, signalled a decisive break with the British tradition of governing within a broad range of political values and social norms. Postwar governments, both Conservative and Labour, largely accepted the goal of full employment, regulation of the economy and a significant role for government in the provision of housing. And while certainly favouring particular constituencies key to electoral success, British national parties implicitly

34 The governance of space in london

accepted the value of communities as key elements of the broader society, neighbourhood identity and cultural diversity.

But Thatcher's election portended not incremental change in political values, and policy manoeuvres designed to manage economic cycles, but a neoliberal revolution, as recounted in Charles Moore's three-volume biography (2013, 2016, 2019). Government in the Thatcher prospectus would be committed to building an internationally competitive economy based on the financial industries, intermediate services and the property sector. Swinging tax cuts on businesses and incomes would, in Thatcher's view, incentivise enterprise and promote growth.

The GLC was viewed pejoratively by Margaret Thatcher's Conservative Government as an essentially regulatory, high-taxation body inimical to the spirit of enterprise in a new era of globalisation shaped by free flows of capital. The abolition of the GLC by the Thatcher government in 1986 marked the end of a century long tradition of managerial governance for London, with a decade and a half interregnum before the establishment of the Greater London Authority in 2000 by Tony Blair's New Labour government

Major industries and traditional labour groups, including mining and unionised manufacturing, were seen by Thatcher as residuals of Britain's past, and were effectively marginalised in her prospectus as inimical to a bright new future based on markets, capital and globalisation. For Thatcher and her government, the logic of transformation included a messaging that Britain was 'first in' the earliest industrialisation era, harkening back to the mid-eighteenth century in northern cities and the Midlands, and would logically be the 'first out' of now-outmoded blue-collar occupations in a marketised, globalising economy.

The power and influence of industrial trade unions, seen historically as legitimate in promoting the interests of workers in wage bargaining and in political action, was now viewed by the Conservative Government as a significant institutional impediment to the new competitive Britain, and would be directly confronted by government.

Margaret Thatcher's agenda and program was centrally concerned with national competitiveness and a larger role for Britain in global markets. But the impacts of government policy were highly localised in cities and urban communities. While the early years of the Conservative government saw vigorous and at times violent protests and upheavals in the industrial communities of the North, the Midlands and Scotland, London and its region were at the heart of the neoliberal agenda. Industrial action in places like Ford's auto plant in Dagenham signalled an intensification of contestation between workers and companies.

The sell-off of council housing promoted by the Conservatives stripped many working-class families of residential tenure. The sale of council flats served to dislodge the working classes, many of whom tended to identify and vote for Labour, while over time creating an owner-occupier class imbued with political values and cultural signifiers more aligned with the Conservatives.

The hallmark project of Thatcherism was the development of the new international financial and business complex at Canary Wharf in Tower Hamlets, in the centre of London's Docklands. The government's partnership with Olympia & York, and with the Reichmann brothers of Toronto, ran roughshod over the protests of Tower Hamlets Council, local residents and workers, while the establishment of the London Docklands Development Corporation (LDDC) represented a new institutional model designed to favour capital over the interests of local government and residents.

Relatedly, the Conservative government's rolling back of local planning powers in Britain significantly diminished the power of local authorities to resist market players and property interests in particular. I address this theme in Chapter 3 (following) as part of the larger financialisation narrative which has transformed so much of London's economy, labour force, social morphology and built environment.

Strategically, the Canary Wharf project stimulated an emergent reorientation of London's development geography. London's corporate sector, including property, financial and marketing concerns, was mobilised in the interests of an emergent eastward development trajectory for the British capital, with traditional districts enlisted as landing sites for capital (Thornley 1992; Travers 2003). New gentrification landscapes emerged within the residential neighbourhoods of East London, in consequence of the weakened tenure of industrial workers experiencing widespread job loss.

Estate agencies and other property concerns established footholds along high streets and other commercial sites in East London boroughs, signalling a more insistent capture of territory, and a wider revalorisation experience in postindustrial London.

Part of the new population inflow to postindustrial East London comprised artists and other cultural actors: an important harbinger of the larger and consequential cultural turn in areas such as Islington, Shoreditch and other sites on the City Fringe. In the early years of Margaret Thatcher's Conservative government, this vanguard included many younger artists, co-operatives and experimental cultural groups.

These initial contingents of artists stimulated a cultural rebranding of the postindustrial landscape, within which they were displaced in turn by more affluent gentrifiers: a process repeated in London as in cities such as New York, Toronto and San Francisco among other examples.

Tony Blair's new labour program: The establishment of the Greater London Authority

The election of Tony Blair's New Labour government in 1997 entailed, among other agendas, a comprehensive reform of London government. The Greater London Authority (GLA) (established 2000) was a signature project of New Labour, a departure from the traditional English socialist traditions of the

36 The governance of space in London

Labour movement, and offering a more entrepreneurial approach to metropolitan governance, oriented toward maximising London's economic potential in a globalising world. While the 32 London boroughs represented the basic level of governance in the 1963 Act, exercising substantial powers of development control, the Greater London Act promulgated by Blair's New Labour Government privileged the GLA as the strategic level of policy making and development.

In particular, provisions in the act for an elected mayor entailed a more entrepreneurial management style than the Leader position in the former GLC. Partly in response to the opportunities afforded London's mayor in the Act, high-profile individuals came forward as candidates, among them Ken Livingstone—who ran successfully against Tony Blair's preferred candidate, and then Boris Johnson, who defeated Livingstone in the latter's bid for a third term as mayor for London. In 2007 the Blair Government made changes to increase the power of the Mayor in dealings with the borough authorities.

Tony Blair's 'New Labour' government tacitly accepted much of the Tory economic agenda, with a stronger interest in 'culture' as growth sector, and an enhanced community development orientation in the new century, as more aligned with Labour values. Conservative Prime Minister David Cameron's agenda included a personal interest in a resurgent East London tech sector with its epicentre in Shoreditch. In each case the central state and its agencies increasingly favoured investment in London, including capital for key institutions and for the technology intensive transportation networks which provided key infrastructural elements of London's economy.

Current mayor Sadiq Khan (since 2016) has continued to develop the economic program for the capital, while introducing new messaging and policies for London, including social and cultural inclusion, affordable housing, public transit and policies for a sustainable London: in effect representing an alternative to the Conservative government in Westminster.

The Greater London Authority represents a major break with the GLC which it replaced after a gap of almost two decades, during which the capital functioned without metropolitan-wide local governance. The GLA and the boroughs still perform regulatory functions. But overall, the imperative of fostering growth through capital investment, higher education, innovation in fields such as digital technology and building design and, importantly, the mobilisation of London's diverse spaces for high-margin redevelopment, represent the leitmotif of governance.

Over the last quarter century, the traditional values and nomenclatures of city planning for London, incorporating commitments to public housing, public services and infrastructure provision, and with a policy emphasis on growth management, development control and regulatory functions, have been supplanted by a narrative of 'regeneration' (Imrie, Lees and Raco 2009).

The origins of regeneration as governance modality and policy approach can be traced to the exigent quality of the policy challenge in response to structural unemployment, deep hardship in traditional working-class neighbourhoods and inequality and poverty experienced in immigrant communities. But regeneration implies a distinctly entrepreneurial model of local governance and planning practice in twenty-first-century London.

Governance for London: State and non-state actors in shaping development

I acknowledge the expansion and reordering of governance in London to take in a far more extensive set of actors, agencies and interests implicated in the reshaping of space than was the case in earlier models. The Mayor exercises considerable agency within the GLA, comprised of statutory powers and the resources of elective office.

London's 32 boroughs are themselves significant institutional actors in the enactment of development agendas, including the implementation of suites of economic development, cultural programs and regeneration schemes, in partnership with market and community actors and agencies. The larger boroughs (including, among others, the City of Westminster, Camden, Islington, Hackney, Kensington and Chelsea, Wandsworth and Southwark) possess sufficient resources and development potential to exercise considerable agency in the international arena of place-making, marketing and capital investment flows.

Other influential actors in the governance of London include variously, coalitions of development interests and lobbyists, regeneration bodies and their constituent representatives and membership, architects, urban design professionals and—of course—political leaders and cabinet members of the British central state. Those generally advocating for development in the metropolis incorporate large numbers of consultants, advisors and lobbyists, including in London as in other globalising cities contingents of international actors, as well as the mainstream media.

These comprise, collectively, powerful agents for growth and change. But London is also characterised by active community and neighbourhood associations committed to having a voice in shaping change at the local level. These include passionate and well-informed groups and individuals, many of whom possess a deep knowledge of London's history, value socio-cultural diversity and connect personally with London's resonant landscapes and built environment.

The balance of power and agency favours capital and redevelopment in London, as the record of the last four decades unequivocally demonstrates. But social actors can be effective in constructing progressive coalitions and in deploying new media to offer alternative narratives in the service of community values. These social groups form a part of a new middle class of knowledge-based postindustrial classes in London which comprise influential agents of community life and neighbourhood formation in the metropolis.

To these resident social groups in London, we can identify as forces for reshaping space in the metropolis key factors including, notably, industrial restructuring

38 The governance of space in london

processes and market shifts in London and across urban systems globally; migration, including international immigration and inflows to London from other British regions; and the reformation of social class, with its impacts on forms of housing tenure and community development.

Conclusion: The changing role of the state and markets in London

I have outlined in this chapter some selective aspects of the production, management and mobilisation of space in London, including reference to signifying changes in the governance of space in the British capital. I have been attentive to important historical moments and episodes over the late modern era, both as signifiers of changing state attitudes toward London, and as links to larger processes (and discourses) of contemporary urban change, within which London occupies an influential niche.

I have incorporated an overview of the changing state attitude toward intermediate service industries as a striking and instructive measure of the policy distance travelled over the past 50 years. The efforts of central government to micro-manage London's commercial office sector in the 1970s, followed by Margaret Thatcher's totemic Canary Wharf financial complex imposed on the postindustrial terrain of the Isle of Dogs a decade later, surely represents one of the most dramatic reversals of state governance and policy practice in urban development values on record.

Of course, London's spatiality extends to districts and communities beyond the City of London and Canary Wharf. In this chapter I described the rise of specialised industries over the nineteenth century, including a distinctive economy of light manufacturing and divisions of specialised labour within the inner northeast, and extending from Kentish Town through the City fringe and Hackney and as far eastward as Limehouse and Millwall. These constituted in many cases communities of deep deprivation and hardship in the postindustrial experience of the 1970s and 1980s. But over the last quarter-century, these districts have been mobilised as sites of cultural regeneration, gentrification and serial reproduction, responding both to larger economic trends and to local opportunity.

The narrative proceeds to a present moment in which London retains its primacy in finance within Europe, but also where rival European cities aspire to hive off segments of London's banking and business cluster in the wake of Boris Johnson's formal declaration of a departure from the single market: a decision opposed by a substantial majority of Londoners in the Brexit referendum in June of 2016.

And it might be useful here to cite a recent paper published online by Nauro Campos and Fabrizio Coricelli which establishes through careful empirical work the relative influence of Margaret Thatcher's neoliberal reform agenda and Britain's entry to Europe as motive forces of growth. They conclude that there is 'little empirical support' for the former, and stronger evidence for the critical turning point being instead 'around 1970 when the UK finally began the process of joining the European Economic Community' ('How EEC membership drove Margaret Thatcher's reforms': Vox CEPR Policy Portal: https:// voxeu.org/article/how-eec-membership-drove-margaret-thatcher's-reforms: 10 March 2017, accessed 15 May 2020).

On the face of it, the Conservative Government's obdurate fixation on Brexit represents both a policy folly as well as an implicit challenge to Neil Brenner's forecast (from the chapter introduction) of a global future within which state power is subject to 'perforation' by lower of orders of government, and in which cities, regions and urban communities can increasingly achieve a larger measure of agency in shaping their futures. But to the extent that the 'perforation' of central government has been achieved in Britain, the record of the last quarter-century discloses a deepening penetration of Westminster and Whitehall not by local and regional government, but by the force of capital and market players.

3

MONEY METROPOLIS

Financialisation, spaces of capital and the property machine

Capital city, city of capital: Introduction

Capital stocks and money flows are important factors in urban development among advanced societies, comprising critical inputs to the financing of systems of production (investments in plant, machinery and labour) and the built environment, as well as enabling the operation of networks of trade and exchange. Salaries and incomes derived from labour support the livelihoods of urban populations, including the consumption of housing, education and myriad quotidian needs.

Capital as a function of urban development in an era of deregulation is also associated with the appreciation of land values and building stock, historically producing experiences of upgrading and dislocations, as David Harvey vividly demonstrated in his monograph on Paris as exemplar of modernity in the second empire (2003; see also Sutcliffe (1973) for an account of the transformation of central Paris in the Haussmann era).

A portion of land value and rents accrue to government and the local state in the form of taxes and fees, enabling the provision of public goods and services essential to social well-being in the city. These operations of public finance represent critical aspects of capital in the city and more particularly the globalising metropolis, providing the means by which the local state can invest in programs which address inequality, notably in housing but also education and health care.

But capital, and more particularly the asymmetrically weighted distribution of money (and monetised assets) accumulating from earnings, rents and inheritance, is also associated with status, privileged access to other resources including property and education and growing socio-economic inequality in the current era of neoliberalism. In this chapter I demonstrate that London represents one of the most instructive case studies in the power of capital to reshape space, place

DOI: 10.4324/9781315312491-3

and territory, as well as contributing to the changing mix (and tenure) of social groups, communities and cultures in the global metropolis.

Following this opening commentary I position London within discourses of political economy, with emphasis on the saliency of the financialisation narrative, leading to an outline of what I describe as territorial capitalism in the London case, and incorporating stimulating commentary from influential scholars in the field. I then offer a succinct outline of the complex interdependency between finance and space in London, including signifying patterns of London's banking and financial geography. This discussion takes in contrasts between the City of London and the Canary Wharf cluster, with respect to institutions, specialisation, design and architectural values, and, relatedly, the production of global imaginaries which comprise features of London's financial ecosystem.

Chapter 3 concludes with presentation of a model of an emergent zonal structure for twenty-first-century London shaped by neoliberal governance, capital, culture and new enterprise formation, effectively enlarging the spatial ambit of impactful development beyond the historic central area, and framing London's distinctive status within the narratives of territorial capitalism: prelude to the more detailed essays on spaces of capital relayering and upgrading in London which follow.

London as centre of capital and finance: Continuity and disjuncture

Finance has been central to London's development trajectory over extended history: as marker of national economic primacy; as enabling platform for England's program of imperial expansion, international trade and rivalry with other European states from the sixteenth century onward; and as qualification of membership within the roster of other European financial centres over history, including Florence, Antwerp, Amsterdam, Paris and Frankfurt.

Finance has also throughout history been implicated in the reproduction of London's internal spaces and territory, including notably the banking establishment in the City of London and affiliated clusters of institutions and social cohorts, and in the operation of property markets. The City of London, together with the extensive docks to the east, comprised the critical spatial platform of mid-Victorian London, as observed in John Darwin's evocative description of the heart of imperial control:

> This was the setting in which London asserted its global supremacy. This was not the London of Whitehall, but the City, the square mile of commerce at the other end of the Strand. Lining its handful of streets, by-ways and alleys were the offices of clearing or high street banks, merchant banks, overseas banks, insurance companies, Indian and South American railway companies, shipping companies and shipbrokers, the

42 Money metropolis

> famous Chinese firm of Jardine Matheson, traders in the mundane (sugar) and the exotic (human hair), great imperial corporations (such as Cecil Rhodes' British South Africa Company on London Wall) as well as a host of highly prized experts like mining engineers. The City's two poles were the Bank of England and the Stock Exchange. Lying nearby were the Docks – the London, Surrey, Limehouse, West India, East India, Millwall and Victoria. Together they symbolized the huge range of its global activities as the centre of world trade, the supplier of credit and the source of foreign investment.
>
> *Darwin (2012: 182–183)*

Flows of capital in the form of investment, loans, deposits and savings and portfolio and pension funds, greatly augmented over the last four decades by global capital, have transformed space and territory over much of London, shaped by the particular imprints of financialisation as process and emergent economic regime.

In the postwar era the expansion of finance and allied intermediate services clusters increased London's national primacy and comprised a key centrepiece of its ranking as 'world city' (Hall 1966), and later represented the key economic specialisations underpinning London's inclusion as one of just three first-order 'global cities' (along with New York and Tokyo) in the last decades of the twentieth century (Sassen 2001[1991]).

The development of a new international financial and business complex at Canary Wharf since the 1980s has augmented London's global city status, at a time of increasing competition from financial centres within East Asia, such as Tokyo, Hong Kong, Singapore, Shanghai and Mumbai as well as capital entrepots like Dubai, and numerous offshore tax havens.

Continuity represents a feature of London's financial specialisation and overall trajectory, in terms of the agglomeration of key institutions (notably the Bank of England and the Stock Exchange), as well as recurrent experiences of crisis, including crashes of stock portfolios, liquidity issues and the crises of existential war. The uniquely privileged governance arrangements of the City of London constituted a principal continuity over the span of London's development history, and enabling other aspects of power projection in the commercial and trading realms.

For much of the City's history a distinctive social order has shaped a particular form of interaction and collective influence. As David Kynaston (1994) has described, the social organisation of London's financial sector, leavened by rich milieu effects and by complementary amenity since establishment of the seventeenth-century coffee house, has constituted a key underpinning aspect of the City's operations. London's strength in banking and finance constitutes a major feature of continuity within England's economy and circuits of trade and exchange, and represents an exceptional case within a backdrop of Britain's economic decline.

Disruptions to London's banking and financial complex have included notably market crashes, company bankruptcies, deep institutional strains owing to currency runs and illiquidity and of course recurrent wars and other conflicts. The speculative nature of finance, investments and loans practiced by institutions and companies in the City of London impart a high degree of volatility to the operational dynamics of enterprise over extended history. That said, the capacity of London's banking and financial establishment to raise capital has constituted a critical factor in the financing of imperial expansion, in the sustaining of trade relations, and in enabling Britain's viability as sovereign state over centuries of war and conflict, including the existential wars of the twentieth century.

London's competitive advantage in banking and finance within Europe will be increasingly tested in the wake of Brexit, observed in the efforts of continental states and business interests to hive off industries, companies and skilled labour. Brexit has imparted a shock to the global perception of the stability of Britain's (and by logical extension London's) status as secure financial jurisdiction, and more particularly the integrity of trust, trading agreements and social relations which support the primacy of banking and finance in London. As London is both Britain's financial centre and a large contributor to central government budgets (and transfers to other regions of the country), the negative implications of uncertainty associated with Brexit extend beyond the capital to the socioeconomic welfare of the UK as a whole.

Financialisation: Political values, property markets and transformations of space

Finance and 'financialization' constitute long-running features of cities, national economies and communities over extended history, which (as Gerald Epstein has observed) incorporates 'the increasing role of financial motives, financial markets, financial actors and financial institutions in the operation of the domestic and international economies' (Epstein 2005 cited in Sawyer 2013: 6).

Crucially financialisation as process is linked causally to neoliberalism (Duménil and Lévy 2005; Helleiner 1994). It was the onset of a new form of neoliberalism ushered in by the election of Margaret Thatcher's Conservative government (1979) in Britain, and then Ronald Reagan as US President in 1980 (who largely followed Thatcher's political script) and ensuing programs of deregulation and marketisation, which enabled a dramatic potentiation of financialisation as policy value and modality.

A corollary process took the form of a political discounting of traditional manufacturing and labour both in Britain and the US, producing large expanses of postindustrial territory available for more profitable redevelopment and capital relayering. Former sites of industrial production and craft specialisation in London have been repurposed over the last four decades for a suite of

44 Money metropolis

higher-value activities, including creative industries, institutions and enterprise associated with the innovation economy, and, increasingly, convivial consumption. In the wake of these landmark political events, London constitutes one of the most compelling demonstrations of the power of financialisation in the spaces of the city.

I turn from a recitation of political factors shaping financialisation of national economies, and a new round of globalisation, to a more grounded consideration of how finance and capital are shaping cities, and more particularly the ordering of space. Shaun French, Andrew Leyshon and Thomas Wainwright assert that financialisation has not been accorded its due influence on the reconfiguration of cities, in the form of upgrading pressures exerted on an increasingly wider expanse of urban space. They refute the notion that space should be 'implicitly subordinated either to the status of mere empirical surface, or to that of an abstract container of financialized capitalism', and instead assert that financialisation 'must instead be understood as a profoundly spatial phenomenon' (French, Leyshon and Wainwright 2011: 800).

While financialisation as applied to spatial reproduction in the city is associated in the first instance with benchmark political developments at the national level in Britain and the US, we can readily trace critical aspects of change which follow Molotch's earlier (1976) exposition of the urban 'growth machine' involving city governments in collaboration with market players and more specifically property firms, acting collectively to more fully realise the profit potential of urban land.

Market actors and representative agencies of the local state effectively collaborate in urban redevelopment, as each party derives significant benefit (rents, fees, higher land value and income for market actors; taxes, employment and other regeneration benefits for public authorities) from redevelopment. With the growing impress of global capital within urban property markets, the local state has evolved from exercising a principally regulatory agency, to entering into forms of partnership with the market in high-value (and high-return) development in the city.

The power of capital in urban growth and change has been greatly potentiated by the deregulation and marketisation policies of the last 30 years. Malcolm Sawyer cites Boyer's contention that the 'financialized growth machine' can be seen as 'the latest candidate for replacing Fordism' and, further, that institutionally 'the financial regime plays the central role that used to be attributed to the wage-labour nexus under Fordism' (Boyer 2000 cited in Sawyer 2013: 15). There may be other contenders for this descriptor in an era of urbanisation in a multipolar world, but Boyer's proposal underscores the pervasive influence of capital in the metropolis.

London constitutes a particularly noteworthy example of this contemporary iteration of financialisation. More particularly a property development complex including banks and other lending institutions, developers, architects, estate agencies, marketing and advertising concerns, construction companies, regeneration agencies and other intermediaries—fuelled by massive flows of off-shore money as well as domestic capital—shapes a distinctive financialisation platform for London's growth machine.

'Worlds apart': Natascha van der Zwan's thesis on polarising effects of financialisation

Scholars working within the field of accumulation regime theory point to the industrial crises of the 1970s, afflicting especially Britain and the US as seminal events in the financialisation storyline. In the US, the Carter administration's failure to attend to critical weaknesses in manufacturing accelerated the shift of capital from the production economy (Bluestone and Harrison 1982) to more profitable opportunities. And as I discussed in Chapter 2 the shift from an economic model incorporating a semblance of balance between marketised services and a unionised industrial economy, to a dominant professional and managerial economic regime, was a defining feature of Margaret Thatcher's policy agenda, widely emulated internationally.

Financialisation operates at different scales within states, economies and society, constituting a far-reaching imperative whose influence certainly includes but extends well beyond the production economy. In a paper published in the *Socio-Economic Review* titled 'Making sense of financialization', Natascha van der Zwan offers an insightful account of the ramifiers of an 'increasingly autonomous realm of global finance' (van der Zwan 2014: 99) within different realms of impact, which in the aggregate comprise a restructuring of advanced political economies and the societies within which they are embedded. Van der Zwan opens with an assertion that 'finance' involves 'not the neutral allocation of capital, but rather an expression of class, a control mechanism … a rationality associated with late twentieth-century capitalism' (van der Zwan: 102), implicitly conducive to serial upgrading in place and successive rounds of dislocation.

Natasha van der Zwan points to the increasing weight assigned to shareholder value as a concomitant feature of financialisation. Two cohorts (or sets of actors within financialisation) are seen as principal beneficiaries: first, 'the past 30 years have witnessed steady increases in dividend payouts and share buy-backs to shareholders' (van der Zwan: 2014: 108); second, various incentive pay schemes have 'enabled top-level managers to enjoy unprecedented degrees of wealth' (van der Zwan 2014: 109).

At the same time this highly skewed distribution of rewards accruing from financialisation has produced downward pressure on workers' wages, although in leading economies of continental Europe these negative effects have been ameliorated to an extent by state welfare and distributional programs, as well as the strength of labour organisation.

Finally, Natascha van der Zwan identifies the 'financialization of the every-day' world as an expression of the cultural impress of finance and money in society. This constitutes a complex field of inquiry, and includes the 'extensification'

46 Money metropolis

of work into the quotidian realm of households, as observed by Helen Jarvis and Andy Pratt (2006) in the San Francisco case. In broad terms the 'everyday' dimension of financialisation penetrates the life of households via the downloading of responsibility of welfare provision from the state to individuals, observed in individual (rather than corporate) contributions to pension funds, and the purchase of medical insurance to guard against debilitating illness, and life insurance: 'an increased convergence of finance and the life cycle' (van der Zwan 2014: 111).

As a contribution to this discourse I later develop a profile of connectivity between financialisation and spatial change in the London case, a pattern with its origins within the business landscapes of Central London but with increasing penetration over a far-more extensive range of territory throughout the capital.

Thierry Theurillat on finance, property and territorial capitalism

Thierry Theurillat contributes to what he terms the 'territorial approach' to investigating relations between finance, the production economy and the city: a model which has deep resonance for studies of contemporary London. He starts with an appreciation of the growth of 'the financialisation of the built environment over the twenty-first-century' (Theurillat 2011: 2). Theurillat asserts that a separation exists between 'economic and financial logics' (4) in the operation of the property market. This tendency, and the risk factors associated with such ventures, generates in turn new demand for 'promoters and intermediaries' in the development of the built environment, 'since they have the essential knowledge of the local markets' (Theurillat 2011; see also Torrance 2009).

These specialists manage transactions which enable 'the transformation of a real asset into a financial asset and the development of a particular investment circuit based on a logic of investment portfolios typical of financial operators' (2011: 5). In London as in other globalising cities specialists are drawn in from architecture, urban design, digital marketing and other place-making fields to re-image (and market) spaces and sites across the metropolis.

Theurillat extends this logic in the form of a model of urban financialisation which comprises the built environment, space and urban territory. Under financialisation the built environment represents a product of capital, which in turn generates the 'diversified infrastructures' which comprise a 'good business atmosphere' conducive to the (re)creation of 'value': an essential instrument for 'promoting growth and development' (7). In larger spatial terms the 'urban territory' comprises a platform 'on which the infrastructures will generate positive effects for the city as a whole' (7): a theme I return to in the concluding section of this chapter, in which I depict the lineaments of a new and enlarged zonal structure for London.

The financialisation of the built environment generates in globalising cities insistent pressure for upgrading and dislocation. In addition to brochures, online press releases and other forms of media promotion deployed by property interests, stories of major projects and associated territorial transformations are routinely published by the media, including free daily papers and online media. These narratives conflate the profit potential of 'property' with London as 'urban territory', permeating the public consciousness, and contributing to London's identity formation and imagery.

To illustrate: during the bonus season within which very large payouts are made to the most successful traders in the City, estate agencies look forward to what is enthusiastically described in publicity pieces as a 'wall of money' hitting the property market in London. This tacit endorsement of the inflationary impact of high earners on property values serves to further marginalise those Londoners not possessing access to wealth.

London as site and exemplar of financialisation

The saliency of finance to London's larger development experience has intensified since the 1980s, an era which Giovanni Arrighi (1994) describes as a 'decade of financialization': foregrounding an era encompassing post-Fordism (and postindustrialism) as process and policy value; see Chapter 2) and a new regime of accumulation among advanced economies and societies. London and New York performed as vanguard cities within which finance assumed the institutional centrepiece of specialisation and competitive advantage in an emergent era of accelerated globalisation, enabled by far-ranging deregulation and marketisation.

Financialisation has been described in trenchant terms by Greta Krippner as 'a pattern of accumulation in which profits accrue primarily through financial channels rather than through trade and commodity production' (Krippner 2005: 174). The singular experiences of London and New York are derived both from the empirical growth of the financial sector including intermediate banking and financial derivatives, as well as from the collapse of manufacturing industries in each case: for London manufacturing accounted for 22 per cent of total employment in 1977, compared to 8.4 per cent in 1996; while the comparable figures for New York were 21.9 and 9 per cent. As a comparative reference, Tokyo, the third 'global city' in Saskia Sassen's influential ranking, manufacturing still accounted for almost 17 per cent of total employment in 1996 (Sassen 1991: Table 8.2: 203).

As observed in Chapter 2, the political values of Margaret Thatcher's Conservative government included both endorsement of a fully marketised economy based on finance, property and business services and a discounting of the value of manufacturing, shaping in turn a narrative of postindustrialism and a profound reordering of the structures of the economy. The residual industrial production sectors in London favour high-value, batch-scale outputs, notably in food and beverages, fashion, household décor and other cultural products, as I depicted in Table 2.1. The imagery of London as global financial capital assiduously promoted by market and state agencies is buttressed by references to the city's culture, connectivity and amenity, as well as London's infrastructure,

48 Money metropolis

skilled workers and universities: complementary features of the master narrative of globalisation.

'Finance' and financialisation in London assuredly means chartered banks, the Bank of England and the stock exchange, and the myriad institutions and enterprises that support London's financial complex. But it means much more. There is the larger narrative of 'making money from money', part and parcel of the global economy of capital circulation, the amassing of funds both legal and extra-legal. Financialisation represents a motive force in growing inequality, expressed in politics, the social structure of cities and communities, and more particularly in the stratification of communities and neighbourhoods in globalising cities. Also attendant in this process is the enormous risk exposure associated with financialisation, certainly pertaining to the privileged holders of large capital pools, but also for pension funds and for individuals, as experienced in the financial crisis of 2008.

Relatedly 'financialization' underpins the restructuring of the world economy, and constituent globalising circuits and systems of cities. Global flows of capital are increasing at a faster rate than flows of goods. These include transfers, government monies and funds, loans, private capital flows and deposits and of course massive flows of illicit money. These money flows touch down within banks and other financial institutions, and within the spaces and territories of cities, (re)producing privilege, polarisation and displacement in urban societies, housing markets and neighbourhoods.

Financialisation is forcibly expressed in the property markets of globalising cities. In terms of the structure of the economy, there is the tight meshing of banks, lenders and other financial institutions with companies in the property sector: land and property management, developers, construction companies and myriad subcontracting companies, architects and urban designers, landscape architects, consultants of various kinds, local government staff involved in project application adjudication, but also regeneration agencies of various kinds.

Banking and finance have been at the centre of London's development trajectory in terms of the structure of the overall economy, and also in geographical terms, occupying privileged space at the core of the metropolis in the City of London. These privileges were safeguarded by the quasi-autonomous political power of the City Corporation within London's governance against serious attempts at reform, until the Greater London Authority Act (1999) which enabled the political primacy of the Greater London Authority within the metropolitan area.

Synergy between the City as semi-autonomous political entity, and the specialised financial institutions domiciled within the Square Mile added to the robustness of its status in the face of pressures for reform, replete with referencing of the importance of the City to Britain's economy and prosperity, as did the cosy social relations between prominent City men and Westminster political figures, sustained through membership in exclusive clubs and societies.

Globalisation and industrialisation produced new ensembles of economic activity over the nineteenth century, including growth in commercial offices, the development of specialised industrial districts of light manufacture in the inner north-east sector of the capital (notably in Clerkenwell and Hackney) and of course the remarkable expansion of the Port of London, described in the preceding chapter. These sectors provided a measure of balance within London's economic structure and expanding space-economy at a time of Britain's pre-eminence as imperial power and trading nation.

The Edwardian era in London saw an increase in commercial functions and services employment (Ensor 1968), while economic development in the interwar period in the metropolis included the complexes of factories, warehousing and distribution in north-west London, including sites in Harlesden, Willesden and Wembley. The 1948 Olympic Games in London shaped new spaces of investment, performance, assembly and social interaction: precursor to the more impactful 2012 London Olympic Games.

Much of London's functional diversity and spatial balance in labour was swept away in the last decades of the twentieth century, first by the collapse of the Port of London in the face of containerisation and changes in seaborne freight rotations, and then by the comprehensive deindustrialisation of the capital's terrains of specialised light manufacturing. Disinvestment was potentiated by the Thatcher Government's thorough political discounting of the value of manufacturing, industrial labour and working class communities over the 1980s (see Charles Moore's biography of Margaret Thatcher, Volumes I (2013) and II (2016) for full accounts of this critical era in Britain's political history).

This industrial restructuring experience, combined with the emergence of a second global-scale financial complex at Canary Wharf, shaped the hyper-specialisation of London's space-economy in favour of intermediate banking and finance, together with the affiliated sectors of property development, consumption, marketing and media.

Second, the imaginaries of capital in London are articulated in many forms of media, including advertising and publicity, in some cases conflating the interests of capital with the general well-being of the metropolitan economy and society. While media coverage of Brexit encompasses in part projected impacts on labour and employment, and communities and households, implications of Brexit for London's banking and financial sector tend to occupy the central narrative of risk associated with Britain's departure from the European Union.

Third, the interests of capital are deeply embodied in politics (and political discourses) at all levels, projecting a unique advocacy for banking and finance. The resentment of London's financial complex throughout much of Britain, derived in part from its unique protections among successive governments and Westminster, as well as the advocacy of the Governor of the Bank of England and other business luminaries, is related in part to the comparative indifference of government to the plight of other industries, including those in the manufacturing sector.

50 Money metropolis

Fourth, financialisation as process has permeated not only political domains and the media, but has also infiltrated social relations more broadly. The vast resources of capital available both to wealthy London residents and transnational elites greatly exacerbates structural inequality associated with massive gaps in incomes, wages and earnings among social groups within the metropolis.

London performs as the city of capital par excellence: a place where money is generated, stored, displayed and flaunted, traded, exchanged and circulated, embedded in the built environment and exhibited as expressions of class signifier and personal success.

Capital and transformations of space and territory in London

Twenty-first-century London has experienced the serial upgrading of sites and the simultaneous reproduction of space and territory over much of the metropolitan area. These fields of upgrading include the principal typologies of space and territory in London, including megaprojects, business districts and clusters, culture, innovation, spaces of spectacle and conviviality and down to the level of neighbourhoods.

Banking, capital and financialisation represent the most potent force in reshaping the morphology and structure of the metropolis. The major institutions of banking and finance occupy prime spaces within London's central area. The very high capital reserves, incomes, profits and salaries associated with finance all combine to offer unrivalled power in commanding the most privileged sites and locations, exemplified by the City of London and Canary Wharf.

Since the 1980s London's banking, financial and business economy has greatly expanded, most spectacularly in the growth of Canary Wharf. The City and Canary Wharf provide London with two distinctive power centres of global finance. They cater to an extent to different markets and clients, and each encompasses specialised services, amenities and territorial ambience, comprising distinctive ecologies of finance, capital and accumulation.

Financialisation also shapes the emergence of production districts and clusters in the city, directly in the dualism of power centres represented by the City and Canary Wharf, but also in the extension of the City situated in the Broadgate, the BNP tower in Marylebone (Figure 3.1) and in landscapes of fund management concerns in Mayfair (Figure 3.2). The growth of Canary Wharf has stimulated property development further east in the former docklands, including the recent high-density residential projects along the Victoria docks.

The intimacy of operational channels between finance, property development and marketing and sales represent a critical feature of the extension of finance into other sectors and spaces of London's economy. Strategically, the development trajectories and space-shaping force of the City of London and Canary Wharf are integral to the (re)production of London's space-economy and metropolitan structure: a theme I return to later in this chapter.

Money metropolis 51

FIGURE 3.1   BNP-Paribas financial complex, Marylebone

FIGURE 3.2   Landscapes of fund management firms, Mayfair

52 Money metropolis

Change and continuity in the City of London

Throughout London's often turbulent history, the City of London has been viewed as a space of continuity and privilege. 'Privilege' as it relates to the City encompasses not just the durability of key institutions, notably the Bank of England, territorial identities and symbolism which project power and influence, and the unique self-governance within the square mile which effectively resisted recurrent attempts at reform; but also the cliques of economic actors and social groups domiciled within the City. Key complements to the City's identity take the form of architectural idioms, the built environment and iconic buildings, including William I's Tower of London, and St Paul's Cathedral, but also the distinctive design ethos of the City's commercial-banking complex.

David Kynaston's (1994) contribution to the discourse on the City of London includes description of the saliency of space and territory as well as the critical importance of agency and institutions. As he recounts the City had established itself as a principal international trading centre by the Tudor era, and potentiated by Thomas Gresham's establishment of the Royal Exchange in 1570. The eighteenth century saw a surge in domestic demand for Asian and American consumer goods (sugar, team coffee, tobacco); a growing European market for the re-export of these commodities; and a protected market in the colonies for British manufacturers, with the City servicing each of these markets.

Coffee houses served as important sites of business, market intelligence and social intercourse in the City, at a time when 'offices were still rare'(Kynaston: 1994: 10). As Kynaston observes, the numerous coffee houses in the City provided working premises where merchants and traders could undertake business, as well as the paperwork associated with transactions, complementing rather than competing with the more formal ambience of the Royal Exchange. The City's coffee houses functioned as places of information sharing and market intelligence, supplying to patrons domestic and foreign newspapers, newssheets, journals, bulletins, customs entry forms, auction notices and price-lists for various commodities and services (Kynaston 1994: 10).

The rapid expansion of international trade and the imperial project in the early Victorian era provided another stimulus to development in the City, and more particularly the emergence of the modern commercial office and property market. As David Kynaston observes, the new head office of the London and Westminster bank in Lothbury (1838) presented 'a complete break with its Georgian surroundings and brashly towering above its next-door neighbour the private bank Jones Lloyd' (Kynaston 1994: 139).

Insurance companies were more consequential than banks in the physical development of the City at this time, with a particular design preference for the 'grand Italian manner' (Kynaston 1994: 139). Over a short period, 1836–1843, insurance companies including the Atlas (in Cheapside), the Globe, the Alliance and the Sun 'were all testimony to a belief in the reassuring properties of uncompromising physical solidity' (Kynaston 1994: 139).

The City's social landscape incorporated clubs and professional associations as well as restaurants and bars, with the latter notionally open to all consumers but in practice presenting a milieu shaped principally by City bankers and traders. The City's built environment was also designed to project an aura of confidence based on tradition and exclusivity, as well as power signifiers which provided the City with a unique status within London's business communities. Together, these internal markers of status, power and privilege represented a bulwark against both commercial competitors and reformist governments.

The City was treated as a special case in the London government review process from 1960 forwards, which resulted in the establishment of the Greater London Council (GLC) in the London Government Act (1963), replacing the London County Council (1888) which had performed admirably. But the decision to leave the City of London Corporation as a quasi-autonomous local government entity was criticised by many advocates for effective reform.

Recasting the City of London: Architecture, innovation, imaginaries

As Maria Kaika has observed the period 1970–1990 represented an 'institutional crisis' for the City, owing in part to threats from successive national governments to constrain its unique privileges as a self-governing body at the heart of the metropolis, even as its financial and corporate office complex buttressed London's global city status. From the 1980s onward the development of the Canary Wharf international finance and business complex situated on the Isle of Dogs presented a competitor to the City for investment in a context of rapid globalisation of capital and money flows.

The challenge of Canary Wharf and competition from leading continental financial centres stimulated response from the City, including a re-imaging process associated with the 2002 Unitary Development Plan. In particular new architectural idioms have been deployed to present a more forward-looking and accessible built form. In place of the legacy of solid (and stolid) office buildings of the nineteenth and twentieth centuries, and including the intimidating perpendicular form of the National Westminster Bank Building, we have whimsical designs with catchy monikers such as the 'Cheesegrater' and 'Gherkin' (Figure 3.3), among other descriptors.

In addition to the changes to the City of London's internal built form and commercial office landscapes, the sequence of architectural experiments in building design and form have produced new vistas of the City as seen from the Thames (Figure 3.4): constituting what Maria Kaika has described as 'an ode to the Corporation's new identity and a visual *coup d'état* against its time-old heritage planning' (Kaika 2010: 453). New architectural values presented opportunity for a corollary shift in the City's institutional gaze from the comforting granite office towers, to an interest in contributing to new landscapes and public imaginaries in a new era of globalisation.

54 Money metropolis

FIGURE 3.3 Reconstructed landscapes and the built environment of the City of London

Architecturally idiosyncratic office buildings are presented by the City officials and commissioned architects as structures which offer more appeal to the non-business public, with a number of them offering tours and opportunities for viewing London's territorial expanse from observation platforms and galleries. These buildings represent a contrast to the generic point towers which make up

FIGURE 3.4 The new face and imagery of the City: A view from the south bank of the Thames

the dominant built form of the commercial office complex of Canary Wharf: inverting standard assumptions associated with the generic skylines of traditional central business districts, and architectural innovation in new corporate complexes in East Asia and the Global South, including the striking office landscapes of places like Shanghai, Singapore and Dubai.

Canary Wharf and Docklands: Capital, culture and territorial change in London

Strategic impacts of financialisation in London have been realised in the remarkable growth of the Canary Wharf global financial, banking and business complex. Canary Wharf represented a profound cultural shift in Britain's business establishment, certainly in terms of architectural values, but also a departure from the cosy traditionalism and insider social relations which characterised the City's banking establishment.

Canary Wharf also produced strategic change in London's territorial structure and space-economy. In the first instance Canary Wharf shaped a duality to the capital's financial structure, establishing a new beachhead for finance and business in the heart of East London. This in turn offered financial and allied business enterprises locational choice within the metropolis,

56 Money metropolis

including access to a financial cluster imbued with an international business culture and global orientation contrasting with that of the traditional milieu of the City of London. The structure and spatial morphology of Canary Wharf includes a roster of major British and international financial and business corporations (Figure 3.5).

Canary Wharf effectively expanded London's metropolitan core into the postindustrial inner city, supplanting a traditional territory of London's light manufacturing economy and associated labour and residential communities, signalling that the needs of the market and more particularly the banking and financial sector took precedence over those of pre-existing communities. Central government brushed aside both the concerns of the Tower Hamlets borough council, which understandably wished to support the interests of the local community over those of the market, as well as objections of the GLC. The GLC had

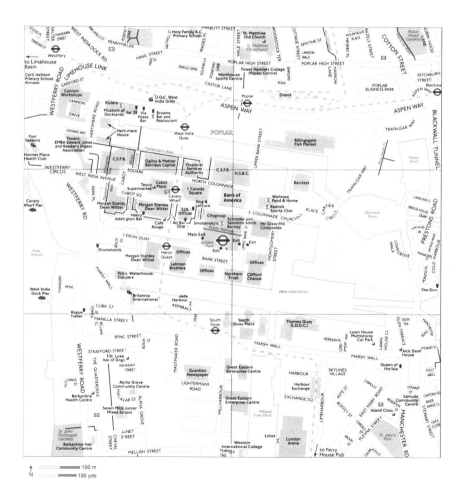

FIGURE 3.5 The corporate landscape of a global financial cluster: Canary Wharf

striven to protect the interests of traditional industry, labour and communities over the extended tenure of the Greater London Development Plan process, as I discussed in the last chapter, until its abolition in 1986.

The development of Canary Wharf signalled the tacit ascendancy of 'space' as principally a resource for redevelopment and upgrading, supplanting the embodied human values of 'place' in the British capital. Neighbourhoods, social traditions and an extended history of municipal socialism in London were relegated to the sidelines of history as outcomes of the Canary Wharf megaproject, affirmed in Margaret Thatcher's famous observation that for her at least 'community' was a concept devoid of meaning. Instead, the values of the market, efficiency and the individual were exalted as the touchstones of a global Britain.

In terms of architectural values and urban design Canary Wharf presents as an isolated high-rise office cluster in the Isle of Dogs, and downstream from the city of London (Figure 3.6). The tallest structures, notably the HSBC Bank, present as exemplars of contemporary high-rise office development, designed to project an imagery of corporate power, although from the ground level within Canary Wharf the built environment presents a more low-key imagery and could pass for CBD streetscapes in Boston or Atlanta (Figure 3.7).

The development of Canary Wharf and the larger Docklands project stimulated successive rounds of redevelopment in East London, reversing the former western orientation of business seen in hedge fund traders in Mayfair, business

FIGURE 3.6 The Canary Wharf International financial complex

FIGURE 3.7 Street-level view of corporate office towers, Canary Wharf

districts in Westminster and the office cluster in Hammersmith as examples. Canary Wharf paved the way for the twenty-first-century reproduction of East London, facilitated by the extension of the Jubilee Line to Stratford, and with the London Olympic Games in 2012 as a further catalyst.

This theme is picked up in Andrew Harris's depiction of ramifications of what he describes as the two defining development models for contemporary London: Canary Wharf as totemic global financial district; and Shoreditch as influential site of culture, innovation and lively social relations (Harris 2014). Each of these districts projects broader influence on the spatial development of the British capital, and each is widely emulated.

Harris acknowledges the irony implicit in the proposal for four 30-storey towers for the Shoreditch area, situated across from a 50-storey tower designed by Norman Foster, figuratively at least putting Shoreditch in the shade; while critics decry the emulation of Shoreditch's model of cultural vernacular across London without reference to the contingent effects of local histories, populations and built form. Proposals for a high-rise condominium curtain adjacent to Shoreditch mimics on a smaller scale the much earlier imposition of a global-scale finance-scape upon the postindustrial territory of the Isle of Dogs: with both projecting the power and imaginaries of financialisation in the British capital

The reproduction of space in London: Shaping an expanded zonal structure

Globalisation has produced a spatial diffusion of capital throughout urban systems, property markets and the built environment since the 1980s, bringing more states and societies within the ambit of neoliberal capitalism beyond the established centres of the North Atlantic realm. Cities such as Mumbai, Shanghai and Seoul have been elevated within global urban hierarchies, and relatedly within globalising territories of banking and financialisation. Defining processes include capital relayering and insistent upgrading, as well as accumulation: elements of transnational elite cohort formation and behaviour, together with the privileges accorded these groups by compliant political agencies.

But London and New York have sustained their positions at the peak of the global financial hierarchy since Sassen's 1991 treatise on global cities, enabled through the proliferation of financial institutions and instruments; through the depth of human and social capital in each city and more particularly skilled labour, including inflows of talent from other cities and societies; and through cosmopolitan cultures (and practices) which pervade both the business environment and metropolitan society as a whole.

In this concluding section I propose to describe the lineaments of an expanded zonal structure for metropolitan London which takes in the spatial coordinates of capital, redevelopment and strategic clusters of the economy: collectively extending London's power projection in Britain, Europe and the global sphere,

Financialisation has served to increase London's remarkable degree of primacy within Britain. London's national dominance in banking, business services and capital projects add to the spatial division of labour problematic famously elucidated by Doreen Massey in 1984, brought up to date in Philip McCann's depiction of 'the UK regional-national economic problem' (2016) represented by London's national primacy. The 'financialisation' problematic within the unitary (but not unified) state of Britain is also manifested in what provincial interests see as a preponderant volume of national revenues allocated to London, observed in government spending on the capital's cultural institutions and hallmark events, with the London 2012 Olympic games, and the massive investment in Crossrail, as cases in point.

Within London the financialisation experience of the past four decades has contributed to the transformation of the metropolis's space-economy and (by extension) metropolitan structure. The traditional bipartite structure of London shaped by banking and financial corporations in the City, and the government, business and cultural clusters of Westminster, has been supplanted by an emergent metropolitan system extending within an axial corridor, and comprising four mega-clusters of high-margin sectors, industries and labour (Figure 3.8)

These comprise: (1) a West London complex encompassing Heathrow and proximate transportation, warehousing and distribution industries; (2) the west-central London clusters situated within the City of Westminster, Camden,

60 Money metropolis

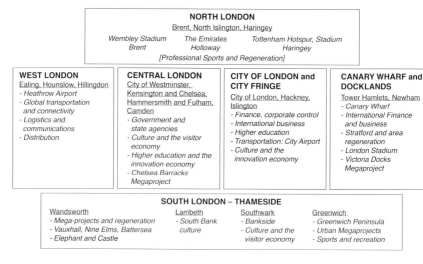

FIGURE 3.8 The zonal structure of twenty-first-century metropolitan London

Kensington and Chelsea and Hammersmith and Fulham, including culture, higher education, the visitor economy and services associated with government and the public sector; (3) The City of London/City Fringe complex of established banking, finance and management specialisations, with Broadgate on the Northern margins and important clusters of cultural, innovation and consumption activities on the City Fringe and Hackney; and (4) a rapidly growing complex in East London anchored by Canary Wharf, and the City Airport, and with a northern extension to Stratford comprising residential, retail, consumption and spectacle. A Southern growth corridor incorporates the O2 stadium complex and the Greenwich Peninsula megaproject, incorporating high-rise residential development with design features by Santiago Calatrava.

This east-west configuration of strategic sectors, propulsive industries and important institutions comprises a global-scale corridor of capital, corporate control, communications and culture, together with the associated patterns of skilled labour including executives, professionals, academics and other specialists in the knowledge and innovation occupations, and supported by lower-wage but essential workers in the final demand services sectors.

The multiple redevelopment narratives that characterise London boroughs situated on the Thames tidewater comprise surely one of the most salient features of the capital's twenty-first-century transformation, extending from Wandsworth and Chelsea in the west to Greenwich, Tower Hamlets and Newham in the eastern sector. While there are complex aspects of change, the defining features take the form of culture-infused projects, notably the massive Vauxhall-Nine Elms-Battersea project in the western sector of this South London-Thameside zone; a central territory encompassing the established South Bank (Lambeth) and Southwark's cultural districts, anchored by the Tate Modern in Bankside;

together with the regeneration of Bermondsey at a lower but nonetheless resonant register of change.

Figure 3.8 also depicts a North London redevelopment zone extending from Brent, through North Islington and Haringey, and with key specialisations in professional sports and local area regeneration programs. The billion-pound stadiums which anchor this zone perform as the focal points of high-density residential neighbourhoods, including student accommodation and local amenity.

The 'effective London region' extends beyond the GLA boundary to include connections between London and Gatwick International Airport, situated just to the south of Greater London in West Sussex, and including dense flows of travellers, supplies of food products and other goods and logistics and communications linkages.

At a larger scale London is situated at the heart of the one of Europe's most strategic development regions, the so-called 'Golden Triangle' including Oxford University (conventionally ranked as the leading university in the world, and less than an hour by rail from Paddington); and Cambridge University, ranked second globally according to the same indices, and under an hour's travel from King's Cross.

More specifically London institutions and companies benefit from particularly dense connections, innovative partnerships and knowledge collaborations with Oxford's world-leading medical complex, and with Cambridge's stellar science institutions, notably the Cavendish Laboratory, and deep capacity for experimental physics among key specialisations. And in 2020 the deep medical/scientific research capacity of Oxford and Imperial College in particular have been mobilised in the quest for an effective Covid-19 vaccine.

Conclusion: Signifiers of capital and spatial reproduction in London

Spatial change in London in the late modern era takes in the (re)production of place and territory, encompassing both the strategic lineaments of an extended zonal structure, just described, as well as complexity and nuance at more localised registers in the British capital. These latter aspects of change form key features of discussion in following chapters, within which I endeavour to tease out the defining complexity of space associated with multiple processes (and narratives) of growth and change in London.

There is also a 'moving tableau' of how scholars structure the defining spaces of change and power projection for London over history. I opened the chapter with John Darwin's evocative description of the City and adjacent Docks as the touchstones of London's imperial projection at the height of empire in the Victorian era.

For Peter Hall, inclusion of London within his roster of 'world cities' (1966) privileged the financial and business sectors concentrated in the City of London, and also took in strategic transportation roles and specialised production industries, while his revisiting of London's power projection in his *Cities in Civilization* (1998) included the City and Canary Wharf but also incorporated culture. Saskia Sassen (1991) included London within her tripartite structure of global cities mostly on the basis of the British capital's platform of banking, finance, corporate control and business services.

As a representative evocation of defining spatial change in twenty-first-century London, I acknowledged in this chapter Andrew Harris's (2014) citing of Canary Wharf and Shoreditch as the most impactful (and widely emulated) territories of signifying redevelopment.

In the following chapter I set out the lineaments of what I identify as elements of the defining spatiality of contemporary London, in the form of the remarkable array of riverine megaprojects extending upstream from the Victoria Dock and Canary Wharf in East London, to the dramatic remaking of space and territory in West London.

This reconfigured zonal structure of London represents the strategic realisation of 'territorial capitalism' in the metropolis, buttressed with important public institutions in the science, higher education, cultural and governance sectors.

A telling measure of the remarkable extent of London's development, and its global reach in the contemporary era is afforded by a comparison of John Darwin's evocative description of the City and extensive London docks at the mid-Victorian height of empire, cited earlier in this chapter, with the expanded zonal dimension of twenty-first-century London presented in this chapter. The City and the docklands still form essential elements of London's twenty-first-century space-economy and global projection, as they did in the Victorian era. But as Figure 3.8 shows 'global London'—where capital touches down, and where power and influence is projected—takes in far more extensive territories than did its progenitors.

What follows in this volume is a sequential explication of the diverse spaces, sites and places that make up the composite forms of territory in London, and which collectively actualise the trajectories of urban change in the present century.

4

GLOBAL CAPITAL AND URBAN MEGAPROJECTS

The appropriation of place-identity and social memory

Capital projects and the globalisation of space in London

I present in this chapter a selective account of the megaproject experience in London: a principal consequence of overarching financialisation processes described in the previous chapter, and comprising a forcible reproduction of space, place and territory in the city.

The contrast between regulatory governance and development control practices of the Greater London Council, London boroughs and the central government through the agency of the Department of the Environment over the 1970s, outlined in Chapter 2, and the openly promotional values of the central and local states in contemporary London, is dramatically realised in the four-decade sequence of megaprojects in the late modern era.

Large-scale redevelopment projects in London exhibit trends observed in other globalising cities. These include the scale of capital deployed; the appropriation of vernacular culture and social memory embodied within design values and marketing initiatives; and the unevenly shared risk between private and public interests. Common to many narratives is a messaging that these large-scale redevelopment projects will enlarge a city's projection within hierarchies of competing jurisdictions while offering local benefits through employment opportunities, the expansion of the residential stock and trickle-down revenues for the state.

The Canary Wharf global financial complex represents the defining expression of Margaret Thatcher's vision of the supremacy of the market and interests of capital over what she saw as the obdurate resistance of local authorities, communities and labour groups to Britain's economic future in a postindustrial setting: an example emulated in many states and cities worldwide. The establishment of the London Docklands Development Corporation (LDDC)

DOI: 10.4324/9781315312491-4

64 Global capital and urban megaprojects

by Michael Heseltine effectively replaced stringent development control at the borough level with an agency committed to fulfilling the unrealised market potential of sites to the east of the City of London.

Canary Wharf served as precursor for a more comprehensive redevelopment program for East London, which has in turn contributed to the production of a reconfigured metropolitan space-economy, described in the preceding chapter. Projects have included notably the Tate Modern (identified as Britain's 'best building' in architectural terms in a public preference survey: The Guardian 2016), and cultural regeneration for adjacent places and territories; the Stratford site of the 2012 Olympic Games; and the Greenwich Peninsula and Victoria Dock development programs.

Megaprojects in this century are framed within local area regeneration programs, in an effort to broaden the base of agency and actors involved, and as a means of capturing public benefits from government investment in infrastructure.

Following this introduction I offer a discussion of the urban megaproject experience as global phenomenon, acknowledging aspects of contingency which shape outcomes on the ground, as a means of contextualising the London case, and including reference to instructive examples. Megaprojects in many cases demonstrate a distinctive aspect of postindustrialism as well as the globalisation of capital, social actors and territory.

I then introduce the megaproject phenomenon in London, and frame discussion within two principal territorial zones as a means of identifying distinctive associations between projects and place which potentiate capital relayering *in situ*. These are situated within the Thames corridor which encompasses a preponderant share of the 'areas of opportunity' within iterations of the London Plan. I offer a description of selective major capital projects extending further down the Thames from Canary Wharf and Docklands, including the Greenwich Peninsula development, and commentary on the extension of redevelopment within the precincts of the Victoria Dock.

Next I discuss projects transforming the Thames embankment upstream from Westminster, with each seeking to mobilise heritage and social history as marketing tools, together with gestures to sustainable development, social inclusion, lifestyle enhancement and employment opportunity. Continuity and contradictions of the Battersea Power Station project are identified in this resonant site: symbolic epicentre of the more extensive Nine Elms-Vauxhall-Battersea development zone.

I also offer commentary on Renzo Piano's Shard tower: a modern icon for London, and demonstrating at one level the supremacy of symbolic representation over use value. Piano's tower lacks the unique intensity of the public connection with the Tate Modern, a few minutes' walk along the Southwark Embankment from the Shard; but benefitted from multi-level state endorsement and public approval in the design phase, and occupies a critical site as lynchpin for a series of cultural-infused developments.

I conclude by identifying the consequences of these capital projects for change within the space-economy and larger zonal structure of London: underscoring the multi-scalar quality and deep impact of redevelopment in the global metropolis.

Megaprojects in context: Values, choices and consequences

Large-scale, capital-intensive, redevelopment projects represent a significant expansion of the projection of urban growth and change in cities and urban networks. Megaprojects are a feature of mature, metropolitan cities within the developed world, as well as capital cities of the Global South, and cities of the Middle East. Apart from the massive scale of financial resources committed to redevelopment, there is also a messaging—implicit or openly stated—concerning underlying policy choices and values.

Sites for megaproject development can encompass a range of place-typologies in metropolitan cities including former manufacturing districts, obsolescent wholesaling areas and goods distribution sites, disused wharves and docklands, underutilised rail station areas and former power stations. London's experience of large-scale redevelopment projects encompasses each of these site categories, suggesting the broad scope of opportunity for redevelopment in the British capital.

Megaprojects and narratives of postindustrialism and globalisation

Projects comprising mostly upscale residential development situated on former production sites signify a demonstration of postindustrialism shaped by political, cultural and social conditions. A decision on the part of the state as development agency not to pursue an industrial regeneration program implicitly discounts the value of jobs, incomes and revenues and diversification of the city and its constituent eco-systems, in favour of a commitment to megaprojects as instrument of global projection.

To illustrate: the Concord Pacific Place megaproject on the north shore of False Creek, an integrated condominium development imposed on a former site of light industry, warehousing and distribution, represents what John Punter (2003) describes as the hallmark achievement of planning and urban design in Vancouver. But another perspective discloses a project imbued with larger meaning and consequences.

The residential trajectory of the Concord Pacific project in Vancouver constituted a political dismissal of possibilities of re-industrialisation for the inner city, and instead endorsement of a talismanic project in support of the City's engagement with the economies and societies of the Asia-Pacific, acknowledged in Kris Olds' monograph on Pacific Rim megaprojects (2001). Hong Kong billionaire Li Ka-Shing was selected among candidates bidding for development rights on the strategic lands on the southern edge of the downtown. The project afforded

66 Global capital and urban megaprojects

opportunity for other Asian investors who, it was hoped, would inject capital and entrepreneurial energy into Vancouver's economy and business sector.

In the event much of the residential 'product' associated with the megaproject was purchased for investment purposes by offshore buyers, continuing Vancouver's extended experience of underdevelopment associated with urban system marginality over its civic history (Hutton 2019). Current efforts on the part of the City to foster an innovation economy within the inner city are constrained by spillover land price inflation generated by high-value housing which effectively limit opportunities for start-up ventures especially.

Seattle, just 200 kilometres to the south of Vancouver on the Interstate 5 highway, presents a contrast to the Vancouver experience in terms of governance values and policy preferences. Overall Seattle's strategic land use choices have favoured retention of industrial land, both for traditional production industries and critical port back-up activities, as well as for the diverse industries and labour associated with the contemporary innovation economy (Barnes and Hutton 2016). Vancouver's inner city presents a far-more coherent urban design sensibility than does Seattle's, reflecting a deep planning culture in Vancouver and its region; but Seattle is characterised by a more robust economy, enterprise structure and labour market.

The global diffusion of urban megaprojects

The lexicon of megaprojects includes mixed-use design, as original template, or in response to changing market conditions. In cases drawn from the international experience of megaprojects, we find high-density housing but also educational and cultural institutions, as in the HafenCity project in Hamburg, representing a more functionally diverse model.

The Minato Mirai redevelopment project on redundant lands in Yokohama's waterfront suffered from falling demand for office development during the deep downturn of the early 1990s. But a broader program including more housing, cultural amenity and public spaces has produced a functional balance of land use. Recent projects in London, notably in the Battersea Power Station development described later, also gesture to the legacy of industrial labour by including spaces for artisanal enterprise as well as mainstream business services.

The Marina Bay comprehensive redevelopment project in Singapore enjoyed the support of government and diverse source investors over several disruptions of the business cycle: a spatial exemplar of the policy values of the developmental state. The Mission Bay project in San Francisco languished through extended periods of governance crises, financing difficulties and public protests concerning the allocation of housing and other benefits.

In the Amsterdam case, policy preferences of the city and the national state have favoured development continuity associated with social needs, including housing and public services provision. The Eastern Docklands redevelopment is characterised by high density housing, and rich cultural and consumption amenity, although the powers of zoning and design allocated to the private developer represents for some an unwelcome departure from the traditional practice in Amsterdam (personal communication: City of Amsterdam director of architecture and design: May 2009).

Urban megaprojects: Acknowledging risks and costs

The prospective scale of profits for corporations engaged in major capital projects, and revenue streams for government agencies, are commensurately large, driven by the typically high price points of property for sale or lease to clients: associated with the steep gradient of upgrading in place, and potentiated by the expansive pools of global capital.

But there are also substantial risks for megaproject developers, and for partner civic agencies, associated with the volume of debt loads, the uncertain cost of borrowing and consumer market response. The pervasive tendency for unit pre-sale in advance of construction is designed in large part to mitigate risk for developers.

This matters in an era where there can be major shifts in business and investment cycles, and in the trajectory of the larger economy. As Bent Flyvberg has demonstrated the scale of benefits are routinely overestimated in the generation of project storylines and imaginaries, while the risks—especially for government and the broader public—are consistently downplayed (Flyvberg 2003).

Both for civic agencies and developers there is also the risk of much larger exogenous shocks, with the 'Asian financial crisis' of 1997 and the deeper and more extensive financial crash of 2008 representing cases in point—and with the 2020-2021 Covid-19 pandemic likely to suppress demand for possibly years to come.

Decisions in favour of redevelopment through the mechanism of megaprojects implicitly discount notions of 'community' as social construct, comprising established neighbourhoods, residents, class values and practices, families and social traditions. In their place are the 'constructed communities' of investors, high-income residents and services catering to these constituencies.

Megaprojects in many cases imply a program of social reconstruction as well as capital relayering and physical redevelopment. They often incorporate strategies of appropriation—cultural, community and social—in the interests of place-remaking, product differentiation and marketing. Signifiers of displaced communities, together with selected aspects or preserved relics of buildings and institutions, are incorporated into contemporary site imaginaries and marketing programs.

The most potent signifier of megaprojects is as multi-faceted marker of globalisation: as aspiration, as trajectory, as rescaling of civic ambitions and intensions and as enlarged field of opportunity, reimaging and competition. Large-scale redevelopment projects also represent a cornerstone of territorial financialisation, discussed in the previous chapter, offering opportunity for accessing ever-larger

68 Global capital and urban megaprojects

pools of capital in the form of investment, taxes, fees and rents. Megaprojects have become key elements of the repertoire of globalising cities—extant, 'in progress', or aspirational—over the era of postindustrialism, marketisation and globalisation which have reshaped space and territory since the 1980s.

London as case study and exemplar

The value of the London case lies in part in the lineage of major projects in the capital, associated with hallmark events, notably the Great Exhibition (1851) in Hyde Park—a progenitor of similar themed events at this scale internationally; and London's three Olympic Games: 1908, 1948 and 2012. The series of capital-intensive projects aligns not only with civic aspiration and access to pools of capital, but also with larger development trajectories for London: as world city and control centre of empire, in the nineteenth century; as one of three first-order global cities in the late twentieth century; and as largest city and financial centre within the European Union—a status which elapsed in 2020 with the formal enactment of Brexit and its isolating consequences.

The construction of Victorian-era civic infrastructure including mainline train stations, power-generating stations and of course the world's first underground system constitute elements of London's rich array of sites with redevelopment potential. The upgrading of rail stations, the abandonment of obsolescent warehousing and distribution facilities, the shift from Thameside power stations to more efficient forms of electricity generation and the collapse of the Port of London with its extensive docklands, warehousing and distribution facilities, have each produced opportunity for large-scale redevelopment.

Where these supply side assets intersect with opportunity for developers in London is manifested in the remarkable array of megaprojects completed, underway or in prospect. From the initial conversions of riverside docks to residential uses, there is now a much larger template of development which takes in redundant or underutilised areas along the River (such as Nine Elms), and sites adjacent to mainline train stations (notably King's Cross).

Capital projects in London draw selectively on the power and resonance of social history, as well as the physical record of industrialisation, labour practices and cultural values associated with redevelopment sites. These themes are invoked by development corporations and associated marketing concerns to capitalise on the localised history (and memory) of place as territorial marker, conferring advantage upon companies operating within competitive markets where there is an increasing volume of 'product' vying for investor interest.

At another level these capital projects imply a rethinking of earlier models of urbanisation, in which younger cities (notably many of those situated in former colonies) performed as sites of active physical development—'cities in the making', in essence—while the principal cities of Europe (Hall 1998) and East Asia (Kim, Douglass, Choe and Ho 1997) capitalised on the heritage values of the historic built environment, notably for the arts, culture, higher education, city branding and of course tourism.

But the collapse of older industrial, warehousing and distribution sites from the 1970s onward have produced new templates of redevelopment in older cities, with London as an instructive exemplar, and with historical imaginaries invoked in high-value regeneration programs.

Urban megaprojects, territorial change and respatialisation in London

In a new era of globalisation post-2008 there is a large market for buyers of high-value residential property, the purchase of which can lead to other privileges of residency for transnational elites. These include the security of investment capital within jurisdictions subject to the rule of law; access to education opportunities (including elite schools and universities in the London region) for family members of international investors; and enjoyment of the broad array of social and cultural amenities available in cosmopolitan cities.

New benchmarks of accumulation achieved by high-net-worth individuals and families internationally have provided enormous pools of investment capital, directed toward residency in high-value projects, encompassing a roster of destination cities such as New York, San Francisco, Sydney and notably London: encompassing opportunities for satisfying social and productive business relations.

The source diversity of capital directed towards London is shaped by multiple factors, including not least the extent and depth of postcolonial relations. London also offers to investors a high standard of international connectivity, and favourable taxation and income reporting regimes.

The emphasis on investment as motive force for 'regeneration' at the borough level tends to favour large capital projects, and the interests of affluent cohorts at the district and neighbourhood scales. As Atkinson, Parker and Burrows (2017) attest in an insightful interrogation of governance in London, the management of space has been effectively reconfigured from stringent development control in support of a balance of social groups, to a far more accommodating approach towards market players and elite interests, reinforcing spaces of privilege across the metropolis.

Cultural capital, social class and territory in London's megaprojects

The London of the first half of the twentieth century was replete with inequality and deprivation over much of its metropolitan area, but presented a semblance of spatial balance in land use, including an economic geography comprised of finance, services, industrial production, warehousing and distribution. While London's social class structure favoured elite populations in terms of environmental quality and amenity, there was provision of space and territory for the

70 Global capital and urban megaprojects

life of the working population of London, including consumption and retail areas across a range of price points; and diverse communities representative of London's distinctive class and occupational structure.

London's experience of restructuring from the 1970s profoundly reshaped its space-economy, employment structure and housing markets, in each case taking the form of upgrading and dislocation. Principal agents of causality have included the primacy of finance, the expansion of higher-order services industries and occupations, the collapse of manufacturing and the privatisation of rental housing, producing steeper gradients of inequality, as described in earlier chapters of this volume, and elucidated notably in Chris Hamnett's *Unequal City: London in the Global Arena* (2003).

Modernist ideas of design for major redevelopment projects in London were realised in the Barbican project on the City fringe—much criticised at the time as a socially alienating form of new brutalism in architectural terms, and following a protracted design phase extending over a quarter-century, with final approval in 1971.

But the Barbican has been reappraised as a striking expression of postwar modernism and progressive thinking about design, as expressed in architecture, housing design and public affinity with place. The Barbican is widely recognised as a vital community space as well as residential community, and encompassing important institutions such as the Museum of London, the Guildhall School of Music and Drama and the City of London School for Girls.

The market appeal of megaprojects includes gestures to local history and social memory, and the iconic features of the built environment—in effect selectively referencing the past, while effacing more troubling aspects of exploitation, deprivation and inequality. Prospective buyers of residential units are invited both to share in the distinctive histories and values embodied in the site (and in some cases a selective preservation of built form), pointing to their cultural sensitivity, while at the same time enjoying the cachet of exclusivity conferred by the project's price points and marketing.

King's Cross: Redevelopment discourse as marker of change in London

The extended saga of the King's Cross project entailed innumerable delays, in a strategic area of the capital bridging Central London and the working class communities of the inner city. King's Cross and St Pancras comprised important elements of the integrated zone of intercity rail services, major stations of the London Underground and warehousing and distribution facilities. Debate over the future of King's Cross exhibits in a particularly trenchant way the remarkable shifts in policy register and public discourse concerning major redevelopment projects in London over the past three decades.

To illustrate I cite here an extract from the 'New Society' section of the *Sunday Times* for 26 March 1989, titled: 'Terminal Illness', and including a punchy and compelling byline: 'the fight is on to stop these homes going for a knockdown price in the name of Channel tunnel progress' (Sunday Times section F: page 1). The basic argument is framed as a form of binary choice between the rights of local householders and residents on the one hand, pending demolition orders to clear the site for terminal redevelopment; and, on the other, those of British Rail, and needs of market interest for fast rail connections between London, other British business centres, and its partner communities (at the time) in the European Union.

Sunday Times reporter Wilfred Peters succinctly states the costs of the execution of a compulsory purchase order designed to clear a 17-acre site for redevelopment by British Rail, as enunciated by Islington Borough Council: the demolition of 61 homes; the loss of 1,300 jobs; and the eradication of 'at least' 15 listed streets and their constituent Victorian houses.

Peters' rehearsal of objections to the demolition order includes lack of consideration of other options for station redevelopment through a less obtrusive tunnelling procedure; projected increases in congestion in an area already experiencing high traffic volumes; inadequate compensation for property owners facing eviction; and negative externalities experienced by residents situated just beyond the perimeter of the redevelopment site.

The value of Peters' story three decades later on includes its sympathetic treatment of the plight of individuals, families and small businesses threatened by eviction through the compulsory purchase order, enlivened by interviews with residents and business owners and operators—offering a window on the intensely localised character of neighbourhood life in King's Cross prior to a totalising demolition and redevelopment process. The texture of description and journalistic discourse include interviews with affected residents and shop owners, and threats of demolition for three pubs at the centre of convivial life in the area: The Bell, the Queen's Arms and the Duke of York.

A telling marker of change in the King's Cross saga concerns the value of heritage in the redevelopment narrative. Here Peters notes in his 1989 reportage that the Victorian Society, a non-governmental organisation with an established record of advising the Department of the Environment on issues of conservation, was the subject of a British Rail attempt to exclude the Society from a list of agencies empowered to comment on the redevelopment of King's Cross (Peters op cit).

The larger template for King's Cross in 2020 and its proximate territory, three decades on from the issues and events just described, includes new institutions, notably the Central St Martins Art and Design College, as well as a lavish profusion of restaurants and bars north of the mainline station (Figure 4.1). Internal station upgrading of infrastructure has been accompanied by a proliferation of high-margin services, shops and amenity located within both King's Cross and the adjacent St Pancras station, as well as impromptu performances by both amateur and well-known musicians which enliven the experience for travellers and other visitors to these iconic terminals.

72 Global capital and urban megaprojects

FIGURE 4.1 Central St Martins and adjacent consumption landscape, King's Cross yards

Centrepieces of the innovation economy are also key features of the latest phase of the redevelopment of King's Cross, notably the British headquarters of Google (situated in a £1 billion 'landscaper' just to the west of the rail terminal), and offices of Facebook, demonstrating the possibilities of major station upgrades as stimulus to other forms of high-value redevelopment and enterprise formation in the globalising metropolis: cumulatively a very far cry indeed from the intensely localised community structure described in Wilfred Peter's reportage three decades ago.

The Thames as development corridor and project site

The saliency of the Thames as critical zone of development (and redevelopment) is reflected in strategic policy and planning storyline in London, from the earliest iterations of programming under the aegis of the Greater London Authority since its establishment in 1990. In this regard Mike Raco cites the narrative of the 'Blue Ribbon' network concept enunciated in GLA's Towards a London Plan document, as follows:

> 'The real heart of London is the river. Look at any satellite image and it is the Thames that dominates. …. it is this huge and beautiful waterway that holds the key to revitalizing the metropolis' (GLA 2002b: iii)

Raco also acknowledges that riverside sites account for no fewer than 22 of the 28 'opportunity areas' described in the 2002 policy document published by the GLA. Later iterations of the London plan have expanded the ambit of opportunity areas within the broader metropolis, but the Thames corridor retains its saliency.

Butler's Wharf in Bermondsey showcases an early example of project-led regeneration in a marginal area of inner London, involving the adaptive reuse of Thameside warehouses and wharfs as upmarket residences, retail and consumption space, providing a lively public realm of convivial experience. Butler's Wharf is also significant for its early demonstration of the reach of redevelopment capital in hitherto peripheral areas of central and inner London.

Since the publication of the GLA's preliminary (2002) development report, more sites and territories have been drawn into the ambit of capital relayering, including a series of megaprojects situated downstream from Canary Wharf. These have cumulatively formed a principal spatial development axis for East London, incorporating projects in the Greenwich Peninsula and the Victoria Dock: each of which comprises constituent communities and neighbourhoods designated in area plans.

More recently a series of large capital redevelopment projects located upstream along the river can be framed as an emergent redevelopment zone for London. These include, notably, the massive Nine Elms-Vauxhall-Battersea project, anchored by the restoration of signifying features of the (former) Battersea power generating station in Wandsworth. Battersea is equivalent in some respects to the image-making qualities of the former Bankside Power Station, shaped institutionally by the Tate Modern Museum and realised both as cultural icon and as marketing hook for upscale housing and amenity provision.

The cultural resonance of the Battersea project underpins its saliency as the centrepiece for a much more extensive local area regeneration program in Wandsworth and Lambeth. Across the river, the Lots Road power station has also been re-imagined and reprogrammed as icon for new residential development along the Fulham-Chelsea Reach, while the Chelsea Barracks megaproject provides a third principal cluster of redevelopment for what I describe as the West London Thameside redevelopment zone.

Megaprojects in Docklands: Capital relayering in East London

The strategic effects of capital relayering in London's Docklands take the form of a continuing sequence of megaprojects downstream from Canary Wharf and the Isle of Dogs. I recall in this regard John Darwin's commentary (opening of Chapter 3) which positioned the docks along with the City's financial institutions as the centrepieces of London's global projection at the high point of empire. As observed in the previous chapter the City continues to perform as London's global financial platform, reconfigured by new imageries and architectural

74 Global capital and urban megaprojects

idioms and infused with sources of capital drawn from East Asia, the Middle East and other regions of accumulation.

We can observe a multifunctional, strategic clustering of development and capital along a strategic corridor with Canary Wharf at its core; a mix of high-density housing, mega-scale retail and consumption at Stratford to the north; and a southern extension in the former Bermondsey-Millwall area comprising the O2 stadium complex, high-density housing within the Greenwich Peninsula and complementary leisure activity. The current redevelopment program encompasses a range of sites from the O2 arena—a legacy of Tony Blair's personal involvement with London as prime minister—downstream toward North Woolwich and Thamesmead West, with a selection of sites shown in Figure 4.2.

The Greenwich Peninsula project, funded by Hong Kong capital, represents both an eastward extension of capital relayering along the Thames as well as an extension of the contemporary globalisation experience of London Docklands. As Figure 4.3 shows, the Greenwich Peninsula encompasses multiple development sites, and presents a distinctive profile of high-rise condominium towers (Figure 4.3), including towers designed by Santiago Calatrava.

The northern section of Thameside and the complex of Royal Docks (including the Victoria Dock, Royal Albert Dock and King George V Dock) comprised principal elements of the world's largest docklands at the height of empire, with the City of London Airport situated between the Royal Albert and King George V docks. The City Airport offers connections between East London and other British and European business centres, while the

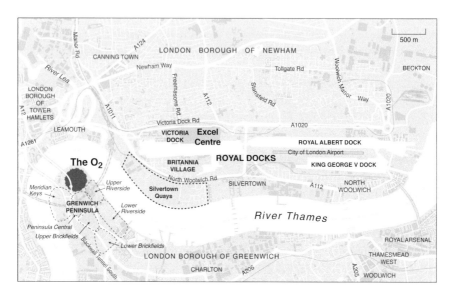

FIGURE 4.2 Megaproject clusters in East London Thameside

Global capital and urban megaprojects 75

FIGURE 4.3 The Greenwich Peninsula project (view from the Isle of Dogs)

Docklands Light Railway brings the Royal Docks within the ambit of the financial and business clusters of Central London. The ExCel Centre functions as a major conference and exhibition complex for the area and was reconfigured to accommodate the Florence Nightingale Hospital during the 2020 Covid-19 crisis.

Downstream from Canary Wharf a sequence of capital projects is transforming the form, imagery and human use of postindustrial space. Examples include the City Island project (Figure 4.4), developed by Ballymore Properties, animated by a marketing tagline that City Island is 'bridging the business might of Canary Wharf with the cultural energy of East London', including the site of the English National Ballet company (Figure 4.4).

The City Island project is situated on the Leamouth Peninsula and will provide for 1,700 residential units and associated amenities.

The Blackwall Reach project takes the form of a new development situated on the site of a 1960s estate, a contemporary upgrading of a postwar residential community and offers 'stunning river or city views' for those interested in participating in a regeneration project with good access to Canary Wharf via the Blackwall DLR station.

The Britannia Village project in the Victoria Dock exemplifies this sequential redevelopment of London's former docklands eastward of Canary Wharf.

FIGURE 4.4 View of the City Island project site

The Village's imaginary is shaped by a range of heritage features, including a former lightship and the last London steamship, mounted on frames on the southern side of the former docks area: reminders of the seaborne transportation vocation of the docks, and also including a large former milling building undergoing rehabilitation (Figure 4.5).

More whimsical features which demonstrate at least the superficial differentiation of imageries across London's megaproject experience include California-style beach bars, waterskiing and 'surfing' opportunities: expressions of the 'new gentrifiers' in the Victoria Dock area which I discuss in Chapter 8.

Renzo Piano's London Shard: A lynchpin of the Thames development corridor

The design and construction of the Renzo Piano's Shard, situated on a strategic site by the river in Southwark, presents a strikingly original architectural design concept, in contrast to the mostly prosaic high-rise office towers situated downstream in Canary Wharf. The Shard represents a spectacular expression of both the larger 'cultural turn' in London's development trajectory and a decisive policy reorientation by Southwark Council and its planning and design staff (Figure 4.6).

Global capital and urban megaprojects 77

FIGURE 4.5 London's maritime history invoked in place-remaking in Docklands

To provide a site for the Shard a serviceable late twentieth-century modernist 25-storey office building at 32 London Bridge Street (completed 1976) and occupied by Price Waterhouse (merged with Coopers Lybrand 1986) was demolished. In interviews I conducted with the Southwark design, planning and regeneration offices 2002–2004, staff disclosed regret concerning the environmental consequences of demolition. But in the words of the (then) Director of Design, Conservation and Archaeology for Southwark, Julie Greer, the Shard's design was 'sufficiently striking and potentially impactful' to merit approval, supported by a high public approval rating exceeding 90 per cent for the project.

The approval of the Shard tower project by a Labour Borough Council signalled a departure in the governance practice and policy values of local political executives and a corollary acceptance of market values as policy signal: an expression of a larger shift in political values and development policy in the globalising metropolis. And here I recall as reference point Peter Hall's commentary on borough politics in his chapter on 'The City of Capitalism Rampant' in *Cities in Civilisation* (1998) concerning the marked radicalism of Southwark borough council in the face of Michal Heseltine's program for redevelopment in East London in the 1980s, relative to the more compliant examples of Newham and Tower Hamlets councils in this period.

78 Global capital and urban megaprojects

FIGURE 4.6 Renzo Piano's 'Shard', Southwark

The approval process for the Shard entailed a complex multi-level governance exercise in design review and adjudication. Following approval of the design by the Borough of Southwark's Planning department in March 2002, and endorsement by the government Planning Inspectorate, Deputy Prime Minister John Prescott signed off on the project in December 2003. Ken Livingstone, Mayor of London, remarked at the time that 'This is a total vindication of the need for a few high-quality tall buildings where they are appropriate and well-designed' (quoted in the London Southeast 1 Community News: London-se1-co-uk/ news/view/732; accessed 20 March 2020).

Since completion of Piano's 1,000 foot-elevation tower a more extensive program of area redevelopment is in progress, encompassing former wholesaling, warehousing and distribution sites and buildings: part of the larger regeneration program in Bermondsey-Southwark. These include plans for a 37-storey tower incorporating housing, retail and leisure uses adjacent to the Shard, and, like the earlier example of the Shard, requiring Borough Council approval for the demolition of a 1980s-era office tower (Southwark News, 31 January 2019). The Shard is also a lynchpin between the Bankside regeneration program and its emblematic Tate Modern Gallery to the west, and the serial redevelopment of the docklands downstream.

The Shard has become an iconic feature of London's landscape and imaginary, arguably the most dramatic new building in twenty-first-century Europe and a reference point for sightlines across much of the city—eclipsing the derivative London Eye Ferris wheel in Lambeth situated across from Westminster. The Shard is not part of the traditional business milieu of the City, almost directly across the Thames, and is essentially self-referential in design terms. But its impact extends beyond the site to encompass a larger profile within the public imaginary.

Megaprojects within west-central London Thameside

While the former docklands sites downstream from Canary Wharf continue to attract capital for redevelopment projects, there is a very extensive program of capital investment along each bank of the Thames upstream from Westminster. These form cumulatively a complex template of project sites and territories at different scales (Figure 4.7), but with each deploying history and heritage in project imaginaries and marketing programs.

As Figure 4.7 shows there is a complex geography of the massive Vauxhall-Nine Elms-Battersea project, with the eastern sector visible from Westminster (Figure 4.8), and with individual projects extending westwards as far as Fulham.

Design aspects of the Nine Elms-Vauxhall project incorporate both heritage features replete with historical allusion, with regard to site and setting; as well as invocation of contemporary signifiers of luxury and exclusive market branding (Figure 4.9) of the properties for sale.

The Battersea megaproject in Vauxhall, financed by Malay interests, represents a spectacular example of twenty-first-century megaproject development. Apart from the enormous scale of the project a defining aspect is the adaptive re-use of the (former) Battersea Power Station, in which a condition of development is the preservation of the four giant chimneys which constitute defining features of the imagery of Battersea. The site offers fast boat connection to

80 Global capital and urban megaprojects

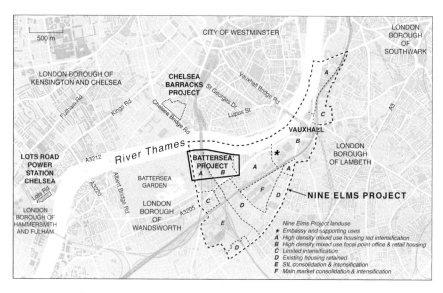

FIGURE 4.7 The template of megaprojects in west-central London Thameside

the City and Canary Wharf, adding to the appeal of the project for highly remunerated financial and professional service workers who represent part of the market constituency of the project.

Across the Chelsea Bridge from Battersea is the Chelsea Barracks project, with Qatari capital as the principal investment source. This project marks the transition of the area from a British army base in the heart of London, to a high-value residential megaproject: a post-imperial turn in the fortunes of this historically privileged district of London.

At a more localised scale the Chelsea Barracks project entailed the eradication of an attractive streetscape of shops and cafes along lower Sloane Street and Chelsea Bridge Road, with no doubt more upscale retail uses to follow as part of the megaproject development, and representing at a lower register an example of capital relayering, upgrading and displacement in the affluent heart of London.

The megaproject experience in London can be effectively read through the perspectives of multiple vantage points: a technique which discloses both commonalities and more importantly/instructively differences. There is a basic storyline; a project history which assists in locating a particular project in the broader lexicon of capital redevelopment projects elsewhere; and more critical perspectives which disclose more discordant narratives and meanings.

Adaptive reuse of disused power-generating stations represents a resonant form of redevelopment in London. The conversion of the former Bankside Power Station in Southwark to the Tate Modern has effectively transformed both the imagery and public use value of this site, while creating spillover

FIGURE 4.8 View of the Nine Elms-Vauxhall site

value for proximate residential towers, other cultural amenities and commercial users. The redevelopment of the station designed by Giles Gilbert Scott was shaped by Swiss architects Jacques Herzog and Pierre de Meuron and was lauded in an article by Oliver Wainwright in The Guardian (2009) as the 'building of the century' in London, derived both from design values and its public appeal.

82 Global capital and urban megaprojects

FIGURE 4.9 Project(ing) market appeal: Branding of project design and exclusivity

Upstream from the former Bankside site two development projects are underway on either side of the river, including the sites of the former Battersea and Lots Road power-generating stations. Each site projects a resonant imagery of industrial London, with massive brick structures and tall chimneys as signature physical and design features. The Lots Road station is situated in the south-west area of Chelsea bordering Fulham and was built by the (former) Brompton and Piccadilly Circus Railway (later part of the Piccadilly Line of the London Underground) in the last years of the nineteenth century as a coal-fired electricity-generating plant, opening in 1905.

At this time substantial housing stock for lower-middle-class residents was constructed in Fulham, but by the 1920s a working class residential trend was well established. With the collapse of London's industrial economy in the 1970s, gentrification was established as social tendency in much of Fulham.

Following the closure of the Lots Road Station in October 2002 the site became the subject of proposals for redevelopment, incorporating the usual mix of high-value residential development and ancillary amenity. Terry Farrell Architects initiated a master-planning process in 1996 for Hong Kong developer Hutchison Whampoa.

The site was rebranded as the 'Chelsea Waterfront' site, and following revisions resulting from submissions from Kensington & Chelsea and a public inquiry, was approved by Deputy Prime Minister John Prescott in 2006. The £1 billion value project was initiated in 2013. The enhanced site valuation and commensurate development was greatly enhanced by its inclusion within the widening ambit of upgrading tendencies in West London, potentiated by a prominent riverside location.

The Chelsea Waterfront project straddles the Chelsea-Fulham border, with the tallest structure, a 37-storey tower on the Fulham side, completed in 2019. Project Formation Architects, in conjunction with Farrells, is undertaking the complex adaptive reuse of the (former) Lots Road Power Station, with completion scheduled for 2025.

Chelsea Waterfront represents a spectacular realisation of the social upgrading of Fulham over the past 100 years, as well as a strategic western extension of riverside megaprojects now extending from Chelsea Creek to the Victoria Dock.

Multilayered contours of the Vauxhall-Nine Elms-Battersea megaproject

The strategic reorientation of development within the metropolis is a critical aspect of the Vauxhall-Nine Elms-Battersea megaproject. The overall situation and siting represent a westward orientation of development on the south bank of the Thames, across the River from Chelsea.

While there are diverse components of the project, situated at different scales, in the aggregate the Vauxhall-Nine Elms–Battersea represents a strategic deepening of capital relayering and redevelopment at the larger metropolitan–regional scale for London. Over the extended development sequence this megaproject zone embodies a set of values, orientations, components and compromises which distinguishes it from its predecessors.

What follows is first a description of the project and its component features, including the conservation of defining physical elements of the Battersea Power Station; as well as design features and heritage imaginaries which underpin project marketing. I then explore contrasts in the project values, goals, design, scoping, expectations and trade-offs as enunciated in key documents issued by the development corporation and local government agencies, informed by selective readings from the research literature as well as my own field work in the area.

Governance and the state: Meta-narratives and the megaproject

The high-level governance vision for the Vauxhall-Nine Elms project is set out in a statement issued in 2012 from the Mayor of London's office, under the imprimatur of the 'Vauxhall Nine Elms Battersea Opportunity Area Planning Framework': one of multiple area plans incorporated in The London Plan. This document includes strategic guidance for linked themes encompassing the planning framework, land use strategy, housing and social infrastructure strategy,

84 Global capital and urban megaprojects

transport, public realm enhancement, tall buildings and environment, as well as technical appendices.

[Then] Mayor of London Boris Johnson's statement of the political view emphasised the economistic dimension of the project, observing that '[t]he area has huge potential to make a significant contribution to London's economy', including, prospectively 16,000 new homes and 20–25,000 jobs, supported by an extension of the Northern Line of the Underground from Kennington to Battersea and Nine Elms. Planning guidance at this level emphasises the economic potential of a new office cluster and high-wage employment, as well as the economic value of high-end residential development, with the latter including a range of tax revenues, company earnings and wages as well as the social benefits of new housing opportunity.

The public consultation process for a project of this magnitude, complexity and prospective impacts was quite limited, extending from November 2009 to March 2010. A principal contrast between the positions of the project proponents and the public was disclosed in the content of the submissions, with the developers asking for high densities to offset the costs of heritage conservation, and with local residents and 'amenity societies' objecting to proposals for very high densities on the grounds of overcrowding, pressures on local services and traffic congestion (GLA 2012b: iv).

A key public policy issue intersecting with the cultural value of the site concerned the public insistence on conserving the basic structure of the Battersea Power Station—with its vast bulk and four tall, iconic chimneys located at each corner of the building—and perhaps constituting even an equally dramatic building form of the (former) Bankside power station which houses the Tate Modern 7 kilometres downstream on the south bank of the Thames (Figure 4.10).

Each looms large in the conscious imagery of the metropolis for many Londoners and visitors alike, but with the consequential difference represented in the public purpose of Bankside for a monumental cultural purpose, in contrast to the Battersea station mobilised principally for place-making for a private sector megaproject. Accordingly much of the debate and discourse between developers, project architects, government agencies and the public centred on precisely how much of the (former) station must be preserved to retain the integrity of design values and historical memory.

Dissecting the Battersea project: Agency, media and reportage perspectives

The essential dualism of values and priorities associated with the Battersea project is framed respectively by the developer and the local state. The former necessarily includes the business case for principals and investors, while articulating in its publicity what it sees as the wider community value of the project based on the iconic cultural value of the (former) station and contributions to the vitality of the public realm.

FIGURE 4.10 View of the (former) Battersea Power Station, Wandsworth

The local state mimics to an extent the developers' storyline concerning the deep public value of the conservation of important features of the former power site, while articulating expectations of public benefit from the project. For projects of this scale the state and market players are to an extent partners, with to be sure different values and constituencies, but joined uneasily in a form of collaborative place-remaking. There are though manifest contradictions, and what follows here is a selective review of more critical perspectives.

Tony Travers, a well-known expert on London planning at the LSE, offers a critique of the scope of public benefits associated with the Battersea project (personal communication 2015). As a mitigating factor Travers notes the very large costs of heritage conservation associated with the Battersea project, including the principal powerhouse building and the iconic chimneys (Figure 4.11). It might be acknowledged though that these heritage features also constitute value for the developers: part of the project-positioning and differentiation the developers have invested in as essential features of the appeal to consumers of residential, office and other business space.

Media reportage forms an important element of the discourse and critique of projects on the scale and prospective impacts of Battersea. A trenchant example is Julia Kollewe's reportage in the online edition of The Guardian, titled 'Battersea is part of a huge building project – but not for Londoners'

86 Global capital and urban megaprojects

FIGURE 4.11 View of the (former) Battersea Power Station chimneys
Photo credit: Lesley Wood

(Saturday 14 February 2015:https://www/theguardian.com/business/2015/feb/battersea-nine-elms-property-development-housing; accessed 4/4/2019). Kollowe's subtitle also conveys the essential subtext: 'the Nine Elms area in south-west London is getting a £15 billion revamp—but its luxury flats, costing £1m-plus, are aimed squarely at wealthy foreign buyers'. She acknowledges the defining features of the site as crucially 'directly across the river from super-swanky Chelsea', and until recently 'wasteland, sheds and warehouses', but now 'one of the largest regeneration projects in Europe, easily eclipsing the £9 billion cost of the 2012 Olympics sites in Stratford' (2/5).

Julia Kollowe's reportage combines a critical edge with resonant terminology, describing what she sees as a 'dramatic transformation' of an area bounded by the river, Battersea Park and the main railway line into Waterloo Station where 20,000 homes are projected as project centrepiece—'most of them luxury apartments in a cluster of high-rise towers that has been dubbed mini-Manhattan or Dubai-on-Thames' (1/5). As context she cites consultancy Property Vision's estimate of 54,000 homes under construction or planned in London as a whole, with the majority priced in excess of £1 million.

Kollowe cites what she describes as an 'upbeat' messaging from Ravi Govindia, (Conservative) leader of Wandsworth Borough Council, acknowledging the high-rise tower configuration of the project overall, but asserting that 'there will be a distinctly London flavor' (2/5). In light of the incomes of Londoners below those of high-earning financial and business professions, Kollowe asserts that 'it is unclear how many of the buyers who get to enjoy that distinctly London flavor will be local people' (2/5).

A subsumed aspect of the megaproject narrative concerns the power of market realities combined with the potency of history, heritage and memory to override political considerations. In this regard Ameeth Vijay offers a nuanced perspective on the power of the 'regeneration' narrative: one imbued with positive social, cultural and ecological ramifiers to mitigate the larger storyline of privilege and exclusivity in the case of the Battersea project. In 'Dissipating the political: Battersea Power Station and the Temporal Aesthetics of Development', Vijay (2018) describes the persuasive value of metaphor, allegory and literary allusions in achieving both political approval and market positioning for the Battersea project.

Vijay opens with an evocative contrast between William Morris's imagining of a Utopian and socialist London, and the concrete realities of the twenty-first-century capital, depicting a Thames returned to a natural state of gardens and trees, described in Morris's *News from Nowhere* (1890), and juxtaposed against a contemporary panorama of glass office and residential towers.

Vijay offers a searching critique of the contradictions of the project through a construct embodying three aesthetic lens: 'the concept of place and place-making, as articulated by the designers'; second, a garden aesthetic and imagery deftly combining 'both a pastoral past and an environmentally sustainable future', which stands for 'a capacious metaphor for the temporal stakes of development and place-making', and, third, the ethos of creativity (Vijay 2018: 611).

Ameeth Vijay acknowledges the value of Davidson and Lees' structuring of megaprojects as prime exemplars of 'new build gentrification', rather than as development model *de novo* (2005). And as Vijay asserts, '[t]he recourse to a highly aestheticised concept of place allows development not only to mediate the structural transformations they are enacting, but also create a narrative and discourse about development that deflects and dissipates political critique'

88 Global capital and urban megaprojects

(Vijay 2018: 611). Recurrent reference in the marketing program to nature, culture and a selective social history serve both to appeal to prospective buyers and to establish a higher purpose for the project than simply a business opportunity.

Vijay's commentary suggests a real-world exemplar of Eric Swyngedouw's commentary (2006) that the power of neoliberalism is expressed in its suppression of the 'political' in public discourses and adjudications of the state, notably in framing urban development decisions.

Relatedly the garden motif embodied in the Battersea project, combined with the appeal of the preservation of design and structural features of the power station redolent of an industrial London, shapes a potent imagery that deflects at least partially the counter-narrative of privilege and exclusivity: a marker of the sophistication of contemporary marketing programs and messaging in the global city.

Chelsea as frontier of megaproject development in West London

Capital projects in Chelsea constitute another phase of redevelopment in London, situated not within the principal zones of postindustrial London, but rather in an affluent territory of the British capital. Here I draw in outline form the profiles of two projects which illustrate different aspects of major redevelopment in Chelsea.

The Chelsea Reach project (Terry Farrell as principal architect) takes the form of a third London power station conservation and adaptive reuse project, following Bankside and Battersea; and new residential towers, adjacent to Hammersmith and Fulham. The Lots Road power station, which supplied power to the London Underground, while not on the scale or cultural resonance of Bankside or Battersea, nonetheless projects an historical identity which can be effectively deployed in place-remaking and marketing [**cover image**].

While the Chelsea Reach-Lots Road Station project represents an extension of capital relayering along the banks of the Thames, the Chelsea Barracks development constitutes a more intrusive scale of redevelopment in this affluent West London borough. Chelsea Barracks occupies certainly one of the prime locations for high-value redevelopment, minutes from the Kings Road and the rich amenities of the area.

The local area's imagery includes a lingering presence of the British army, including not just the former barracks which constitute the principal site of redevelopment, but also the well-known Chelsea Hospital and pensioners' residency located across the street from the project.

To create the street frontage for this project, an attractive and charming array of galleries, art shops, cafes and restaurants was demolished: hardly a social calamity on the scale of the displacement of former working class communities effaced (and histories selectively appropriated) by developers along other sites along the Thames Corridor; but a regrettable loss of local amenity, ambience and character nonetheless (Figure 4.12).

The Chelsea Barracks project is funded by Qatari interests, and represents a broadening in the global base of capital available for large-scale redevelopment

FIGURE 4.12   Chelsea Barracks street-level hoarding, Lower Sloane Street

in London. The process of achieving approval from Kensington and Chelsea borough council was complex, and included many revisions to project scale and amenity provision. It must be noted however that these processes of revision and resubmission are common to megaproject experiences generally and, even following reductions of scale and provision of greater public amenity, still leave generous margins for profit.

If the upscale quality of the site and adjacent spaces represents a distinguishing feature of the Chelsea Barracks project, and signalling a much higher register of capital relayering, another locational feature takes the form of its siting almost directly across the Thames from the Battersea Project: an iconic element of the massive Nine Elms-Vauxhall project described earlier in this chapter, and signalling a westward extension of Thameside megaprojects.

Conclusion: Multivalent impacts of Thameside megaprojects

I have elected here to describe a series of capital projects which represent both continuity and novelty in two principal development zones of the riverine metropolis: the long-running storyline of the extended Docklands area, initiated by the totemic Canary Wharf international finance and business complex in the 1980s; and, second, projects centred on the repurposing of disused but high-resonance power stations in Wandsworth and Chelsea.

I've inserted the London Shard as a key component of the larger trajectory of redevelopment along the Thames, in part owing to its dramatic effect on London's skyline, and its contribution to innovation in architecture and urban design, as well as its impact on the local regeneration narrative. In this regard the Renzo Piano Building Workshop (RPBW) is engaged with partners in the development of a new office block and ancillary retail at the northern end of Bermondsey Street. The Shard, like the Eiffel Tower, is primarily a striking exemplar of symbolic representation over use value, but is also invoked as talisman for projects both within the local precinct and along extended redevelopment corridors.

The projects described in this chapter cumulatively represent spectacular expressions of the 'territorial financialisation' process discussed in the preceding chapter, and with increasing source diversity of capital associated both with project development and with unit sales. These constituencies include investors domiciled in former British colonies, exemplified by Hong Kong, Singapore, India, the Middle East and the US: suggesting a contemporary variant of postcolonialism situated at the upper echelons of wealth, class privilege and aspiration.

It might be argued that the extensive patterns of large redevelopment projects situated along the banks of the Thames constitute a massive layering of inert capital situated within the former docks and industrial districts of London, mostly owned by offshore interests with limited interest in the daily life of the metropolis. But the semiotic value of these territories is carried forward from the Victorian age of empire to the present era. While the Docklands (together with the City of London) comprised an essential spatial feature of London's imperial platform and power projection in the nineteenth century, as observed by John Darwin (Chapter 3), it now performs as both symbolic and concrete manifestation of globalisation in place.

Relatedly the four decades of megaproject development along the banks of the Thames generate important aspects of *spectacle*. This is certainly the case for affluent investors who purchase (and have privileged access to) space within both the new-build residences and renovated heritage structures along the River— vantage points emphasised within the promotional materials circulated among targeted groups in marketing campaigns launched by property interests.

But there is also at least a selective public aspect of spectacle and social experience to acknowledge in connection with London's megaprojects. These features are observed in the array of restaurants, bars and other consumption amenities associated with the King's Cross redevelopment. Within the more dramatic landscapes of the Thames, the adaptive reuse of former riverside power stations lends themselves to a more explicit aspect of public spectacle, realised in the community connection with the Tate Modern in Bankside, and in the preservation and exhibition of heritage ships in the Victoria Dock.

In the exclusive Battersea residential project there is a crafted aspect of spectacle to acknowledge, as it represents a critical element of the marketing program, along with allusions to nature, community and literary history. The restored power station structure itself presents an impactful site imaginary, and incorporates the usual array of lively consumption outlets available to residents and visitors.

Global capital and urban megaprojects 91

There is also a selective conservation and display of industrial artefacts associated with the (former) power station for public view, artfully curated and attractively displayed (Figure 4.13). Visitors are invited to share in the public appreciation of the considerable heritage value of the former power station, even while ownership and residency within the living spaces is largely restricted to affluent populations.

FIGURE 4.13 Public display of curated industrial artefacts, Battersea project site

5

CULTURE AND THE REMAKING OF PLACE AND TERRITORY

Space, local regeneration and the instrumental use of culture

Introduction: Spaces of culture and cultures of place in the global city

To some urbanists London presents as a functionalist metropolis, in contrast to the deep urban cultures of the continent: a roster which includes cities such as Paris, Barcelona, Rome, Florence, Budapest, Haarlem, Brussels and Vienna, among others. In each of these exemplary cases, urban identity, and symbolic representations of city and community have been shaped by experimentation in the visual and performance arts, by signifying cultural movements, and by the expressive power of the built environment.

In his compendium *Cities in Civilization* (2000) Peter Hall depicts cities such as Paris and Vienna as critical wellsprings of cultural innovation, while London is given credit for the stagecraft innovations of Shakespeare, but is accorded a more functional label overall as the 'utilitarian city' of infrastructural innovation in the late modern era.

In addition to the influential roles London has played in the innovation of metropolitan transportation systems and urban infrastructure, its identity has been formed (and performed) by the exercise of power relations: as capital of a highly centralised unitary state; as global centre of banking and finance; as metropole of empire and colonisation; and as counter-hegemon to the principal continental powers and their capitals since the seventeenth century. 'Power' in London is expressed in key institutions—political, administrative, ecclesiastical, military, ceremonial and financial and corporate—and in the resonant symbolism embodied in places and buildings.

The forceful influence of these power complexes should not, however, occlude the saliency of London in the history of the arts and design, and more particularly the emergence of a global scale cultural economy in the late twentieth century,

DOI: 10.4324/9781315312491-5

documented in the publication of the Creative Industries Mapping Document published by the UK Department for Culture, Media and Sport in 2001. This influential study set out the UK's specialisations and strengths across a range of cultural industries, but also served to underscore London's dominant position in culture and the arts nationally.

The storyline includes London's deep and extended history of cultural influence in literature, the stage, film, music and book publishing industries and the Festival of Britain in London's South Bank in 1951. There are also important connections between the arts, culture and politics, as elucidated in Astrid Proll's edited volume on *Goodbye to London: Radical Art and Politics in the 70s* (Hatje Cantz, 2011). Then there are the cultural institution centrepieces of London's economy of spectacle, encounter and experience to acknowledge. But in this chapter, I am principally concerned with a depiction of London's cultural industries and creative enterprise and labour, played out within the spaces and territories of the capital.

There is of course room for individual agency in the arts. But I argue that culture in the city is produced, co-produced and reproduced through complex interaction fostered by what I see as the critical framing matrices of cultural formation, organisation and expression: social; institutional; aesthetic; territorial; and political-administrative. I include as example the remarkable representation of cultural institutions in Camden, which contribute to the formation of the space-economy and overall morphology of the borough.

Next, I offer an overview of a sample of London districts and communities which offer insight into the representation of cultural industries at the local level, including pressures for upgrading and the reproduction of space and territory in the late modern era: a departure from the durable constructs of industry, production systems and labour over history. I conclude Chapter 5 with a case study of Bermondsey: a compelling storyline of the often-conflicting interests of artists, creative enterprise and the market and more particularly the ever-present property sector: each of which promotes in various ways the serial upgrading of place and territory, and constituting a hallmark of contemporary urbanism.

Social actors in the production of culture and creative space in London

Culture in its various forms represents an important influence on the spatiality and territorial identity of the city. For many cities there is a tension between orthogenetic cultures, expressed in landscape features, architectural idioms, iconic buildings and social norms, and the pressures exerted by modernisation, explicated in an influential text on *Culture and the City in East Asia* (OUP 1997), co-edited by Won Bae Kim, Mike Douglass, S. C. Choe, and Kong Chong Ho. These tensions can be managed, mediated or repressed by the state. In western societies and cities, there are structures designed to commemorate experiences of

94 Culture and the remaking of place and territory

nation-building, but are increasingly subject to contestation and reinterpretation by social groups.

Expressions of culture embedded within London's built environment include the design motifs of buildings, including Westminster Abbey and Wren's St Paul's Cathedral in the City. Major government buildings in Westminster, including the House of Parliament and numerous state agencies in Whitehall, embody the cultures of power in a unitary state, expressed through scale, clustering, resonant design values and strategic location.

In the nineteenth history, the development of mainline train stations and the Underground produced impactful forms of industrial architecture, as well as station design embodying important stylistic motifs, including Art Deco features (East Finchley), the futurist form of Southwark Station on the Jubilee Line and George Gilbert Scott's gothic front for St Pancras.

Throughout London's history, the social ordering of space projected a sharply hierarchical form, including at its peak the lavish residences of the London aristocracy and members of the royal family, and a much more numerous contingent of the wealthiest representatives of finance, commerce and trade, the latter comprising a particular form of accumulation through mercantilism.

Other significant cohorts included members of the professional class, dominated by barristers, solicitors, accountants and medical specialists; highly skilled designers, artists and artisans—each of which cohort exercising a measure of influence on the cultural form of the city; and then a heterogeneous pool of merchants, builders and intermediaries.

These groups exercised varying degrees of influence, in places of creative enterprise and work, and also in the cultures of the city as expressed in social identity and everyday consumption preference. Other influential social cohorts were shaped by immigration flows which have transformed districts and neighbourhoods throughout history—each of which has contributed to London's rich palette of cultural values and creative practices—ineluctably salient features of London's cosmopolitan culture and economic dynamism over time.

London's emergence as both the world's largest port, and Britain's largest industrial city in the nineteenth century, produced a development trajectory and rescaling of space and territory within East London and along the River and Thames estuary. These processes in turn are associated with the production of working-class culture in London, together with distinctive community formation, and with lived experiences including deep deprivation and inequality, as well as social, communal and religious observance. Industrial disinvestment contributed to a pervasive evacuation of working-class labour and households in London over the 1970s and 1980s, indirectly creating the territorial basis for artists and other cultural actors.

Culture is not solely a fixed construct within the metropolis, reflecting the hierarchy of institutions and the social order within places of work and residential communities, but is also represented by spaces of flows, and temporary markets and other retail spaces, and patterns of circulation in the metropolis.

New entrants to communities of artists, designers and creative enterprise stimulate innovation in cultural performance. Market players are implicated as principal agents in the reproduction of cultural places, including professional and commercial companies, most forcefully represented by the property sector.

Culture and the geography of the metropolis: Situating London

The geography of culture in the metropolis incorporates structural elements: these take the form of clusters of production, consumption and exhibition arrayed within space and territory. There is a powerful interdependency to acknowledge here: patterns of culture and creative expression in the city vividly demonstrate Edward Soja's injunction concerning the reciprocal relations of 'space-shaping industry' as well as 'industry-shaping space' (Soja 2000: 166).

While some artists prefer to operate as individuals within discrete spaces, many others have contributed to the formation of cultural districts and neighbourhoods in London, and seek to co-locate with other cultural workers to secure productive forms of collective practice and identity formation.

The cultural sector and associated patterns of residency and social life reflect larger structures of hierarchy and privilege, typified by the permanence of principal cultural institutions, and the incomes of elite cultural professionals which afford a measure of personal agency within London's housing market. The more precarious lives of many artists striving to secure living, practice and exhibition space in the city forms a trenchant social narrative.

At the larger scale of the metropolitan economy, a study sponsored by the Greater London Authority, authored by Lara Togni, observed that 'historical and market forces have led to the formation of a world-leading arts, culture and creative industries cluster, constructed on a base of global institutions, knowhow, financiers and distributors, and education facilities .... London "specialises" in culture and the creative industries', with growth fostered through specialisation, clustering and spillover effects (Greater London Authority/L. Togni 2015: 62). As a measure of London's preeminence within Britain, London encompasses 55 per cent of the UK's film companies, and 67 per cent of the jobs in film and video production' (op cit.).

In London the scale and resonance of cultural institutions are influential in the formation of territorial identity, notably in South Kensington and Trafalgar Square among other examples. In other respects, the emergence of cultural spaces represents an evocation of the industrial district, encompassing patterns of co-location and systems of exchange enabled by specialised divisions of labour. In London there are established clusters of cultural industries and creative labour, notably in Camden, Hackney, Islington, the City of London and the City of Westminster, the latter including Soho, described in one report as 'one of the world's most influential media clusters, specifically of film, advertising, TV and radio firms' (Greater London Authority/L. Togni 2015: 51).

96 Culture and the remaking of place and territory

Peter Hall's (2000; 2006) novel reconfiguring of the global city repertoire and spatial representation to include culture and tourism as well as power and influence contributes to how we think about the foundations of primacy, building on the earlier focus on banking and corporate head offices which characterised the principal discourses of the late-twentieth century, articulated notably by John Friedmann (1986) and Saskia Sassen (2001[1991]).

London exhibits representative aspects of the relations between space and culture, and territory and creativity, owing in large part to scale and specialisation, and to the larger narratives of capital relayering, upgrading and dislocations played out across the metropolis. In this interpretation creative spaces of the city represent fluid terrains subject to infiltration by capital and higher-margin enterprises, enabling episodes of restructuring and the appropriation of space.

Since the 1960s and 1970s, districts and territories in London have been brought into play as spaces of creative expression, as well as performing as locus of co-habitation, social exchange and conviviality. The legacies of industrial disinvestment and decline in London include a devastating loss of livelihood and working-class community life, but another residual takes the form of a unique landscape for recolonisation by artists and gentrifiers within the professional and managerial classes.

European cities have developed comprehensive redevelopment programs within extensive postindustrial terrains which include 'culture', realised within project design idioms, architectural innovation, key institutions and enterprise formation. These include notably Berlin (Kreuzburg, Prenzlauer Berg), Hamburg (HafenCity), Milan (Bicocca), Barcelona (Poblenou), Manchester (Ancoats) and Leeds (the Textile District).

London has the largest extent of postindustrial territory of any European city, including not just the inner north-east districts and the extensive docklands from Butler's Wharf to the estuary, but also areas situated within the industrial estates of north-west London, notably Willesden, Park Royal, Wembley and Kilburn—each the subject of local regeneration programs. There are traces of earlier practices of skilled artisanal labour, notably in Clerkenwell and Shoreditch, which are carried forward to systems of creative enterprise in the twenty-first century.

London ranks among the leading global cities in terms of major museums, galleries and exhibition space, along with Paris and Washington DC. Cultural institutions are well-established in central districts such as South Kensington, Bloomsbury and Bankside, augmented by a continuing program of new institutions as well as major upgrades to existing ones across the metropolis.

These structural features of the cultural economy are counterposed against more fluid aspects of London's arts scene. The fortunes of some areas experience decline while emergent sites arise within the media and the public imagination. Artists seek areas which embody authenticity over those which have become more commercialised and increasingly attuned to consumer markets. Clusters of artists and galleries also contribute to dynamics of upgrading, as cultural imaginaries are enlisted by developers in the branding and marketing of new commercial and residential projects.

The layering of culture in London: Community, society and space

The industrial city was defined in large part by factories and warehousing infrastructure, and systems of distribution, together with allied labour, social cohorts and communities; while the postindustrial city was dominated by the high-rise offices of the central business district, and by the residential preferences and cultural signifiers of an ascendant new middle class.

The cultural economy of the city comprises important institutions and infrastructures, including museums and galleries, and film production and post-production, but is also shaped by diverse and nuanced developmental conditions. Communities of artists, professional design firms and, increasingly, recombinant industries and firms which effectively mesh culture, technology and consumption are fostered by diverse development conditions and factors, including local regeneration programs. What follows are descriptive sketches of the principal matrices of culture and creative enterprise in London.

The social matrix of creativity: Community ethnographies and practices

Much of London's creative dynamism is derived from the quality of social capital at the community level. At this local scale culture is shaped by a layering of histories of immigration and settlement, by class and identity, and by household formation and affiliated social practices.

Elements of social structures over history can be traced through residuals of class, physical landmarks and traditions of work, production and consumption. But as consequences of London's deep history of immigration, employment outcomes of industrial restructuring and the transnational cultures characteristic of globalising cities, many of London's communities have been comprehensively remixed in social terms within the postwar era.

Signal events have included the end of empire and postwar immigration from former colonies—and more recently involving a wider range of source states and societies; the traumatic socioeconomic effects of industrial restructuring from the 1970s onward; and the comprehensive shift from a public sector rental housing market to an increasingly private housing market in London.

There are formal representations of London's multicultural populations, including cultural expression associated with places of community assembly, as well as special events designed to celebrate the richness of creative practice on the part of specific ethno-cultural groups. Freedom of worship, including processions and professions of faith in public places, as well as ritual conducted within

98 Culture and the remaking of place and territory

mosques, churches and synagogues, reflect not simply the acceptance of diversity but an affirmation of its value in a global metropolis.

There are also secular expressions of community cultures observed more informally within the streets and spaces of the city, as elucidated by Timothy Shortell in his monograph on 'everyday globalization' (2014), with observations drawn from field study in Brooklyn and Paris, and which I return to in a discussion of vernacular space in Chapter 9.

Enhancing the creative energy of London at the local level are informal events organised and hosted by community and neighbourhood associations, ranging from impromptu local musical performance and theatrical offerings, to block parties—each of which adds to the cultural capital of the metropolis. Some of these are supported by borough councils or non-governmental organisations in recognition of the value of local cultures to the social vibrancy of the city.

There are though forces which undermine the cultural diversity and vibrancy of London's communities and neighbourhoods, not least including the dislocations shaped by capital relayering and gentrification. There is a broad pattern of displacement associated with upgrading, including of course the sustained record of gentrification experienced in the former working-class communities of inner London from the 1970s onward, but also now including the infiltration of 'super-gentrifiers' who possess the means to bid up prices and rents in traditional upper-class communities such as Highgate, Chelsea and Richmond.

Relatedly, London, like New York, San Francisco and Vancouver, experiences what has become termed the 'empty house' syndrome, in which foreign and other non-local buyers purchase homes in the city principally as investment, or as possible hedge against political persecution within states and societies of origin. At a certain threshold, this tendency can become sufficiently pervasive as to produce tears in the social fabric and patterns of connectivity upon which much of the city's cultural vitality depends.

The institutional matrix of culture in London

The centrepieces of culture in London include the exceptional concentration of major museums, galleries and exhibition space: London has the largest number of these critical cultural institutions, with Washington DC and Paris rounding out the top three ('Top spot in world museums chart shared by London and Washington'; The Guardian; https://www.theguardian.com/culture/2017/jun/05/top-spot-in-world-museums-shared-by-london-and-washington).

Major London museums and galleries include the British Museum (in Bloomsbury), the Tate Modern (Bankside), National Gallery (Trafalgar Square), Tate Britain (Pimlico), the Royal Academy in Burlington Place, Piccadilly and the cluster of institutions in South Kensington, notably the Natural History Museum, Science Museum and the Victoria and Albert Museum.

Each of these attract visitors drawn from the continent and overseas as well as domestic sources, and benefit from donations from government, from supporting

Culture and the remaking of place and territory 99

societies and from individual patrons. London's cultural infrastructure also includes its extensive array of drama and musical associations, companies and institutions, as well as performance spaces, including the Royal Albert Hall, and over 50 playhouses.

London's principal cultural institutions are cornerstones of the visitor economy, and provide opportunities for leading artists, while many offer programs for younger artists, in the form of education and training, curatorial experience and opportunities for exhibition. London also encompasses numerous art schools, both public and private, and apprenticeship programs offered by institutions and companies in the cultural realm, including architecture, urban design and fashions.

Like other global cities of culture and creativity, London is endowed with educational institutions which specialise in the arts and creative professions, including the Royal College of Music in South Kensington (Figure 5.1), and the University of the Arts London (Figure 5.2), situated in Pimlico, close to the Tate Britain gallery.

In addition to specialist institutions offering professional education and training for aspiring artists, many of London's universities and colleges offer degree programs, courses and co-op opportunities for aspiring artists and other creative workers, for example, in the visual and performing arts, architecture, urban design, archaeology and creative writing. The Greater London Authority (GLA)

FIGURE 5.1 The Royal College of Music, South Kensington

100 Culture and the remaking of place and territory

FIGURE 5.2 The University of the Arts London, Pimlico

is also a major sponsor of the arts in the capital, as are the London boroughs: many of these have placed the arts and culture at the centre of local regeneration programs.

Community cultural development and expression is fostered by a host of institutions which offer resources for artists, a place of communal expression and identity formation and supports for marketing and sales. There is a particularly rich mixture of institutions in London, which include not just the principal galleries, studios and co-operative organisations, as important as each is, but also sites of cultural (and increasingly intercultural) organisation and expression. These include community centres, places of spiritual observation, ad hoc agencies and institutions and a rich matrix of non-governmental organisations.

While narratives of culture in London tend to privilege central boroughs and districts, there are many examples of creative institutions and initiatives throughout the wider metropolis. An example is the 'White House', described in The Guardian as 'Dagenham's experimental art factory', and prospectively a 'great suburb-wide experiment' in community based cultural regeneration, as reported by Stuart Jeffries in 2016). Jeffries describes a sample of activities which include 'a bling soft play centre, an OAP brewery, and a socialist Jack and the Beanstalk' (Jeffries 2016: 1) (https://www.theguardian.com/artand-design/2016/nov/29/gentrification-dagenham/white-house-create/jeffries; accessed 27 June 2018).

In many of London's districts and communities, there is a proliferation of 'third spaces', such as coffee houses and cafes, and also retail markets, exemplified by Spitalfields and Camden Lock, which afford opportunities for cultural expression and self-actualisation. These sites can also offer possibilities for outreach and marketing. Most include notice-boards, contact information for artists and other cultural actors, groups and consortia, and, in some cases, opportunity for performance on-site—scheduled or impromptu: part and parcel of the assemblage of culture in the modern metropolis.

Other institutions can in part be temporarily repurposed for cultural activity in the form of impromptu performances, adding nuance and diversity to London's cultural landscape. These include the mainline rail stations which offer lively opportunities for artist-audience interaction, including musical performers.

The aesthetic matrix of culture: Landscapes, memory and the built environment

For many artists and other cultural actors, the quality of place, including the complex and contested semiotics of landscapes, territories and local histories represents not just background features, but rather critical stimuli to imagination and creativity.

For some artists these places constitute an agreeably convivial atmosphere for creative enterprise, with prompts and cues embedded in local landscapes, resonant buildings and attractive space for work: complements to the consumption amenities that typically abound in cultural production zones. We can trace a progression from the earlier stages of cultural makeovers of postindustrial districts from the 1970s and 1980s, to the more textured recasting of place and insistent upgrading experiences of the twenty-first century, within which cultural production is increasingly under pressure from high-margin consumption amenities.

102 Culture and the remaking of place and territory

Complex narratives of place, including contestation, struggle and inequality, form essential complements to personal cultural identity and creative aspiration. For these individuals the idea of creativity takes the form not simply of a response to agreeable landscape features of imagined histories of place, but rather in part at least a response to the complex social histories and political contestation redolent of urban history. I cite here Murray Mckenzie's study (2013) of Victory Square in Vancouver: in the present an aestheticised heritage site attracting artists, and a fulcrum between the downtown and the gentrifying Downtown Eastside; but historically a place of racial prejudice, poverty and class struggle.

London abounds in places imbued in the histories of urban crises which find expression in many cities but are arguably more deeply part of London's record of forceful place-making. These include the sites of serial immigration, relayering and displacement (exemplified by Spitalfields); working class districts characterised by inequality and struggle chronicled by Dickens and Marx (Clerkenwell) and more recently the social implications of industrial collapse over the 1970s and 1980s, exemplified by the social histories of Shoreditch and Southwark.

The territorial matrix of creativity: Cultural production districts and zones

The growth and development of culture at the localised scale finds spatial expression in heterogeneous spaces of expression, production (and co-production), consumption and experience. These include the rich spaces of 'cultural quarters', at the district or neighbourhood scale, as Graeme Evans has described in his influential text on cultural planning (2001), and have included integrated zones of cultural production which in important respects exhibit features of the classic industrial district.

London presents a rich case for the study of cultural industry districts, owing in part to examples of film, music and television production in places like Shepperton and Abbey Road. Overall, cultural mapping of London discloses representations of creative activity throughout the metropolis, including the neighbourhood level vibrancy co-produced by immigrants, students and visitors—each of which contributes to the vitality of the city.

At the same time, there is a discernible swath of creative activity extending across London north of the Thames which represents a structuralist feature of 'culture' within the metropolis, comprised of major institutions, cultural districts, clusters of firms within the cultural sector (such as advertising, media and fashion), galleries and salons and workshops and studios.

Since the 1980s a sequence of cultural production district formation constituted an important spatial expression of change in London. These typically comprised features which suggest analogues of the classic industrial district of the manufacturing age in advanced societies, including: specialised industries; integrated production systems and localised network formation; spatial, social

Culture and the remaking of place and territory **103**

and technical divisions of labour; structures and building types deployed for production; and systems of affiliated services, notably communication, management and distribution.

London represents a particularly resonant case, with New York, Chicago, Berlin and Shanghai as other examples, of cities encompassing cultural industry districts, with principal cases including Clerkenwell, Shoreditch and Bermondsey Street. Industrial history is a key to developing an understanding of the specialised industries of each, gesturing to continuities in localised divisions of skilled labour involved in cultural production in each case.

What's different in the present era, though, is the more rapid sequence of restructuring of industries, enterprise and labour within London's cultural industry districts, shaped by capital relayering, property revalorisation and upgrading, and the role of technology deployed in cultural production processes which enable sourcing and marketing across space.

The political—administrative matrix of culture: The local state, creative spaces and regeneration

Cities encompass diverse landscapes, communities and built environments, contributing to a cultural identity (or plural identities) shaped by patronage, design values and in many cases everyday practices of residents and groups. The cultural geography of London, as for New York, Paris and other global cities exhibits complexity: in the values, behaviours and practices of residents and visitors; in the housing choices and consumption preferences of workers across industries and classes; in the stylistic reference points embodied within the built environment; and in the layered histories of place and their signifiers.

There are established cultural signifiers, as represented in places of worship and public assembly, memorials and in places of intensive social flows, as well as industries and institutions. In influential cases there is sufficient representation of culture in terms of artists and cultural actors, signifying landmarks and institutions to form identity at the local administrative level. These include notably the quatrième arrondissement in Paris, Bicocca in Milan, Fremont in Seattle and Hollywood in Los Angeles.

In these cases creative enterprise is officially recognised by state agencies as central to local identity and the economy, supported by suites of supportive policies and programs. Supportive programs for the arts, culture and design in London include at the metropolitan level major festivals include those celebrating (and promoting) design, exemplifed by the annual Design Festival (Figure 5.3).

London boroughs support aspirational programs for 'culture' within their respective areas, as opportunity for self-actualisation and employment opportunity for residents, and as place-differentiating features for marketing purposes. 'Culture' and creativity is recognised and operationalised in multilevel governance, in the form of supports for cultural institutions, public events and festivals, performance space and education.

FIGURE 5.3 'Design takes London': Event signage at the Victoria & Albert Museum

Critical processes of upgrading, succession and dislocation

Many of London's cultural industry spaces are shaped by an upgrading of cultural enterprise, from initial artists and craft workers, to more profitable cultural enterprises and highly remunerated professional labour. Supply and demand factors complicit in the overall upgrading tendency include in the case of the former contingents of creative professionals, artists and curators (full-time professionals, as well as many part-time workers, volunteer docents and the like), augmented by graduates from London's many art and design schools and cultural programs.

Second, the rise of an 'economy of innovation' shaped by more intensive applications of technology in experimentation and production represents a forceful upgrading tendency, with culture deployed more as agent of milieu-formation than for its intrinsic value.

A third factor of dislocation takes the form of the inflationary influence of London's property market at play in many of these districts, as investors, developers and estate agencies seek to capitalise the higher rents accruing from upgrading. 'Culture' is frequently brought into the representation of sites and properties as a crucial signifier of value.

Fourth, the spatial expansion of London's prime sites for redevelopment inevitably brings in territories previously seen as 'marginal' in the metropolitan context to the mainstream vortex of property markets, especially, but not exclusively, in East London. Thus the market reconfiguration of Shoreditch as situated on the 'City Fringe' by planners and policy-makers from derelict postindustrial to an 'artists pastoral' (Harris 2012) introduced new residents as well as cultural performers, effectively shifting the gentrification frontier eastward into Dalston.

Finally, culture underpins a consumption orientation within London, exemplified by the cafes, restaurants and wine bars that increasingly infiltrate the cultural production zones of places like Shoreditch, Bermondsey, Clerkenwell and Soho.

The territorial ambit of culture in London

I now turn to discussion of instructive case studies—necessarily selective and drawn from a large field of possibilities. The cases below offer opportunity for insight into the intimate relations between space, culture and territorial change in the global city, at different registers of scale, innovation and disrupture. I start with Camden, which incorporates significant clusters of business and professional services and important medical research activity, but also exceptional representation of cultural institutions.

I offer an overview of the experiences of Clerkenwell, a district with a rich history of immigration, settlement and cultural change in London over two centuries.

Then I present an account of connections between historical legacies, landscape features, conservation and upgrading in Bermondsey, situated within Southwark, an instructive site of cultural renewal. In each case we can readily trace interdependency between the growth of cultural industries and serial upgrading in London.

Camden: Morphology of a cultural archipelago

Each of London's 32 boroughs comprises sites and landscapes of culture, incorporating diverse social vernaculars, agencies and institutions. The London Borough of Camden incorporates in its southern districts especially concentrations of business, professional, educational and state agencies: important elements of the platform of governance, agency and management which comprises the power projection of Central London.

Camden also exhibits a particularly rich mixture of cultural production and exhibition within its diverse spaces. Overall, there are informal features of culture in the north of the borough, including the well-known Camden Lock Market, popular with locals and visitors alike, and with a more formal institutional quality proceeding southwards.

Camden's cultural morphology and landscape of institutions extending north-south (Figure 5.4), starting with the Chalk Farm Underground station, incorporating the Camden Market, Jewish Museum and the David Roberts Arts

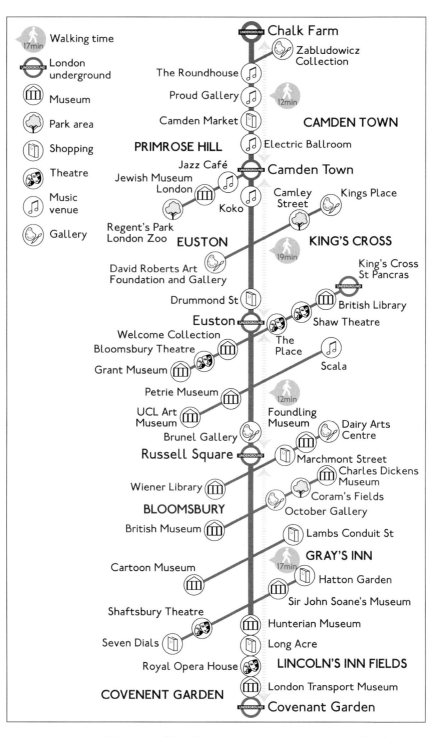

FIGURE 5.4 Cultural line map of Camden: Lineaments of an institutional archipelago

Source: Camden Creative

Foundation and Gallery. The Euston-St Pancras and King's Cross area, a critical transportation interchange which enhances public access to Camden's cultural institutions and sites of creative enterprise, includes the British Library, the Wellcome Collection, Bloomsbury Theatre and the Shaw Theatre.

Continuing southwards Bloomsbury's cultural landscape is anchored by the Russell Square underground station on the Piccadilly Line, offering convenient access for visitors to the British Museum, the Charles Dickens Museum, Coram's Fields, the Shaftesbury Theatre and Sir John Soane's Museum.

Finally, the southern quadrant of Camden's cultural archipelago comprises the Covent Garden public market, the Royal Opera House, the London Transport Museum, Long Acre and the Hunterian Museum. In sum Camden encompasses an exceptionally rich cultural landscape of the arts, creative production and consumption: comprising in the aggregate a centrepiece of London's cultural economy.

Culture and the reproduction of space in Clerkenwell

Clerkenwell constitutes an instructive territory within London, shaped by a rich mixture of precision industries, social history and influential polemical treatments, While a number of districts have been identified as encompassing defining qualities of London as metropolis, this claim is perhaps more convincing for Clerkenwell than most as it incorporates finance and business services, skilled industry and batch industrial production and artisanal labour, consumption and distribution as well as housing estates.

Figure 5.5 shows the complex spatial dimensions of Clerkenwell, shaped by a distinctive pattern of streets, land use and the built environment. Field survey work I conducted in Clerkenwell (multiple site visits 1998–2006) disclosed a bifurcated structure within the district. The northern sector incorporated residential and retail land use, including the Exmouth market and high-quality public housing in the Finsbury Ward; and with diverse industries and commercial activity south of Bowling Green Road.

Craft industries and trades have constituted a critical platform for Clerkenwell for two centuries, shaping a distinctive form of path dependency and high-value production through the tumult of wars, social conflict and the dislocations of industrial restructuring. My visits to Clerkenwell in the waning years of the last century disclosed important residuals of the area's specific industrial history, including workshops and light industry, and new commercial and professional offices, reflecting the larger services orientation of London.

This pattern of activity corresponds to field work I conducted in San Francisco's South of Market Area in the 1990s and early twenty-first century, where I observed residuals of the machine shops and allied labour specialisation that comprised important features of the Bay Area's industrial sector; but where digital enterprise, commercial offices and consumption industries were clearly in ascendancy in the early years of the new century (Hutton 2008: 178–221).

108 Culture and the remaking of place and territory

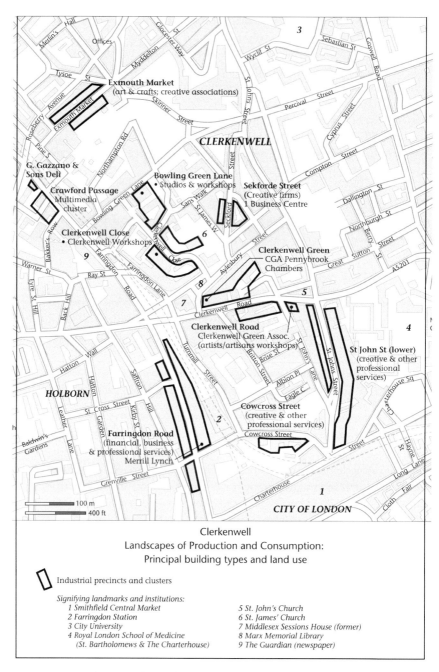

FIGURE 5.5 Landscapes of specialised production and consumption: Clerkenwell c. 2005

Clerkenwell comprised an important element of the arts-based cultural turn in London. Field research yielded observations concerning the growing presence of artists, art supply shops and galleries within the spaces of Clerkenwell, as well as the conversion of former artisanal spaces and workshops for arts collectives, with the Clerkenwell Workshops as a striking example. My visits in 2003 and 2006 included interviews with young artists attracted by the resonant built environment, the quality of space and affordable rents, as well as the opportunity to exchange ideas with other artists.

This phase presented as one of vibrancy and experimentation; but in time gave way to a cycle of upgrading and displacement, as agents of the property market identified opportunity for mobilising both the distinctive quality of the built environment and the artistic ambience for attracting new occupants (and higher rents). The Clerkenwell Workshops transitioned to a creative setting for architects, interior designers and other professionals, including upscale consumption opportunities on site, notably a bistro and bar, and has experienced multiple sequences of internal upgrading and new tenant profiles over the last decade.

The upgrading experience in Clerkenwell included studios and workshops arrayed along Bowling Green Lane, exemplified by the adaptive reuse of an industrial building at 1 Sekforde Street. Following its conversion from workshop to seniors' community services centre, the building was purchased by Nick Mason and Rick Wright of Pink Floyd, and rebranded as 'Britannia Row 2': a high-amenity habitus for elite creatives, following the installation of the original Britannia Row studio in Islington. Creatives, including musicians, designers and other artists, could avail themselves of an organic latte bar, massage room and a 'chill-out' facility: emblematic of the high point of Clerkenwell's cultural makeover in the first decade of the present century.

This exuberant era of cultural regeneration in Clerkenwell has given way to a mainstream business profile, including in the southern precincts especially high-end consumption, including hipster barber shops and regional Italian restaurants (Figure 5.6). This consumption platform, as well as an increased presence of business service firms and professional offices, is likely to be part of the area regeneration program associated with the impending opening of the Farringdon station of the Crossrail/Elizabeth Line system.

Bermondsey and the interdependencies of the arts, regeneration and upgrading

Bermondsey offers an instructive case study in the instrumental uses of culture in local regeneration. Its compact territory (bounded by the Thames on the north, Borough High Street on the west, Southwark Park on the east and Old Kent Road on the south), and visually resonant brick warehouses, has served as terrain for successive episodes of upgrading, triggering dislocations of working-class residents and low-income tenants in the larger area.

FIGURE 5.6 'Apulia' Pugliese restaurant, Cowcross Street: Clerkenwell

Bermondsey has experienced a sequence of upgrading processes over the past quarter-century, including the entry of artists and other cultural actors and institutions as agents of change. Bermondsey's physical distance from the earlier terrains of artists' movements and initial gentrifiers on the City Fringe afforded the area some measure of insulation from the effects of upgrading. But there is clear evidence that the property market is now the major player in the area's growth and change.

An extended program of field work conducted in Bermondsey since 2003 has disclosed a familiar and instructive pattern of culture-led regeneration. I was originally drawn to the site as a prospective field of study on the south bank of the River, as a counterpoint to the larger territories of arts, culture and creative enterprise in Camden and South Kensington in Central London and Clerkenwell and Shoreditch on the City Fringe.

I engaged in a program of cultural mapping, interviews and site photography over the past 15 years, attentive both to aspects of change and continuity in the area. I was also interested in the role of institutions and programs in the shaping of arts-based regeneration programs, allied with the conservation of historic buildings which served as key infrastructures for artists and other creatives, shaped by a distinctive linear configuration (Figure 5.7).

Impetus for the regeneration of Bermondsey Street and its environs included its designation by English Heritage as a Conservation Area, leading to a restoration of the warehouses and older structures, including the historic Leather Market (1833) on nearby Weston Street. As Pevsner and Cherry noted in their volume on *London South* (Yale University Press: 1983), the most architecturally coherent buildings were arrayed along the southern end of the street, with the 'best house' in the area a late-seventeenth-century house 'with a pretty oriel window and double overhang' Cherry and Pevsner 1983: 608).

Bermondsey Street: Reports from the field, 2003–2018

My initial visits to the area circa 2003–2005 revealed a large contingent of artists and other cultural actors and institutions, including representations of each domiciled within the nineteenth-century (former) warehouses along Bermondsey Street and adjacent streets. Bermondsey Street originally served as high street connecting the riverside with the parish church at the southern end of the street, and had experienced a form of industrialisation in the nineteenth century as warehouse and distribution centre, specialising in spices and other foodstuffs and leather goods. These distribution functions experienced first a slow decline, then a near-total collapse owing to the dramatic contraction of the Port of London over the 1960s and 1970s.

The Ticino Bakery at the southern end of Bermondsey Street represented a vestigial presence of the area's industrial past, founded by the Gianelli family, the latest of generations of family bakeries, and among the seven oldest bakeries in London, according to the *Townfish* food guide (published 11 February 2013). The Ticino Bakery specialised in long-fermentation sourdough bread and included signage (in 2003), warning residents and passersby not to impede the movement of trucks along this already gentrifying corridor.

In this early stage of Bermondsey's art-led regeneration, contingents of artists, architects and other design practices were complemented by larger enterprises and institutions. Among these was the Delfina Studio Trust, established in 1988 in Stratford. In 1997 the Trust relocated to an attractive brick structure at the

112 Culture and the remaking of place and territory

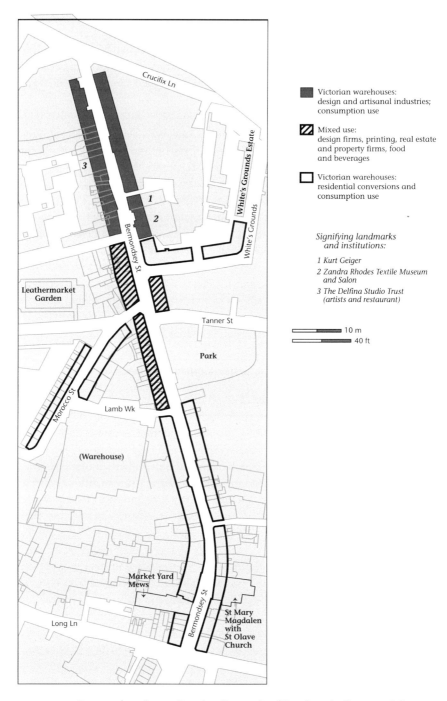

FIGURE 5.7 Bermondsey Street, London Borough of Southwark: Structural features

Source: Hutton (2008)

northern end of Bermondsey Street, on the site of a former chocolate factory. For a decade Delfina served as an attractive domicile for resident artists, including many from the continent, and the program included annual public exhibitions.

As noted in Delfina's online information sheet, 'exhibitions were mainly dedicated to medium scale, solo projects within an emphasis on site specific works and installations by both British and international artists', and thus 'contributing to the larger regeneration of the area' (http://www.delfina.org.uk/exh/index.html; accessed 27 March 2019). The artistic mission of the Delfina Trust was advanced by the exhibition of works by resident artists in Delfina's on-site café, which also attracted patronage from other artists and designers in the area, and contributing to the conviviality of this highly aestheticised landscape.

During this phase of Bermondsey's cultural regeneration, other principal enterprises included Kurt Geiger, a successful shoe design firm (with its initial space situated in Bond Street in 1963) with annual sales exceeding £100 million, located at 75 Bermondsey Street, and, next door, the Zandra Rhodes textile museum. The latter project entailed a partnership between Ricardo Legoretta, an internationally acclaimed architect, and local architect Alan Camp. The provision of flats at the back of the project provided a critical cross-subsidy for the viability of the project.

The recent turn in Bermondsey Street's arts and culture trajectory can be viewed through the lens of two significant institutional manoeuvers: the first represented by the closure of the Delfina Trust studios, gallery and restaurant in 2006, replaced by the Delfina Foundation for the support of young artists; and the recent event taking the form of the opening of Jeremy Jopling's White Cube Bermondsey Gallery in 2012.

A signifying feature of the latter was Jopling's closure in 2012 of the White Cube gallery in Hoxton Square—a hallmark institution of Hoxton's cultural turn from the 1990s, and implicitly signifying a territorial shift in the fortunes of the two areas, with Hoxton and the larger Shoreditch area increasingly drawn within the ambit of the innovation economy of inner north-east London, as I describe in Chapter 7 of this volume.

White Cube Bermondsey occupies a large site on the west side of Bermondsey Street, situated on a former warehousing operation dating from the 1970s just as East London's industrial decline was underway, and represents the dominant institutional presence in the street (Figure 5.8). White Cube Bermondsey comprises three principal galleries, an auditorium, bookshop and warehousing capacity.

Bermondsey Street: Culture, property and the aesthetics of upgrading

The larger Southwark borough area which encompasses Bermondsey encompasses an important territory for arts-based redevelopment, including major institutions such as the Tate Modern and the Globe Theatre. But increasingly, property market players have set a new agenda for redevelopment in the area where developers capitalise on the cultural branding of the area.

114 Culture and the remaking of place and territory

FIGURE 5.8 The White Cube gallery: Bermondsey Street

The proximity of new housing to cultural institutions has produced experiences of conflict, highlighted by complaints from investors in a condominium tower proximate to the Tate Modern that visitors to the Tate can look directly into their premises: a putative case of invasion of privacy. This charge was dismissed by the courts, incorporating a reasonable judgment *prima facie* that the opening of the Tate preceded the construction of the residential tower.

Like other areas in London characterised by a territorially bounded character, resonant social and industrial history and landscapes suitable for conversion to higher-value uses, Bermondsey has been effectively rebranded by developers and estate agencies. Signifiers of an aesthetic turn in the long-running upgrading of Bermondsey Street in this century include site-specific refinishing of heritage buildings to appeal to more affluent tenants and clients (Figure 5.9)

As particularly ebullient messaging from property concern, Knight Frank Rutley underscores Bermondsey's arrival among the capital's most attractive residential areas, proclaiming Bermondsey as putatively 'the best place to live' in London, shaped by its location, heritage landscapes and convivial consumption opportunities.

Among those who KFR see as principal markets for residency in Bermondsey include 'investors and professionals', and more particularly 'City workers who enjoy being able to walk or cycle to work', as well as families seeking housing for their children (https://www.knightfrank.co.uk/blog/2018/08/20/Bermondsey-rising-ten-reasons-why; accessed 26 March 2019).

Culture and the remaking of place and territory   **115**

FIGURE 5.9 Aestheticised building exteriors within heritage sites: Bermondsey Street

Reportage both from local advocacy groups and mainstream sources offer a less rosy view of property market trends in the area. Dave Hill contributed to an online issue of the Guardian UK, an essay on the high-end residential turn in Southwark titled 'Regenerating Southwark: urban renewal prompts social cleaning fears' (Hill 2014). He opens with an affirmation of Southwark as 'prime site for the capital's housing crisis rage', ranging from 'unsold £50 million apartments at the top of the Shard to the redevelopment of postwar municipal estates' (page 1 of 5).

Hill quotes Peter John, Labour leader of Southwark Council, as advocating for a mix of high-quality new housing in the borough, 'from council houses to homes for millionaires' (page 2 of 5). Southwark Council's overall development plan incorporates a 'jobs and growth agenda', entailing partnerships between Council and property developers, with the prospect of low-income housing accruing from planning agreements with development companies.

But Hill worries about the prospects of delivering on the promise of local regeneration, in view of Southwark's social divides, exemplified by the well-maintained White's Grounds council estate presenting a 'telling contrast to the reclaimed Bermondsey Street with its White Cube art gallery, stylish restaurants and boutique hotel': a micro-scale example of a larger dichotomous representation of twenty-first century London between 'the glistening, gentrifying new London that sucks in global wealth and an older, wearier one that seems at risk of being devoured' (page 2 of 5) (https://www.theguardian.com/society/2014/oct07/southwark-london-regeneration-urban-renewal-social-cleaning-fears; accessed on-line 27 March 2019).

Establishing a direct line of causality between artists, cultural practice and serial redevelopment episodes requires a careful inquiry into the experiences of specific areas and sites. But it seems clear from this current exposition on Bermondsey that the interplay of forces simulates a familiar dialectic, within which an initial inflow of artists-led are implicated in the regeneration of a compact space, stimulating interest on the part of property interests seeking a marketing hook which differentiates properties within larger portfolios of housing stock in the metropolis.

And in this study area, the process has almost come full circle: if the initial pioneering artists to Bermondsey Street were drawn to the area by an attractive mix of low rents and an industrial built environment carrying the imprimatur of heritage status, and in time serving as an element of market positioning by local estate agencies; then we can also recognise the impact of a new population of more affluent residents on the configuration of local businesses.

On return site visits to Bermondsey Street and its environs in the autumns of 2016 and 2018, what I observed was a clear trajectory favouring high-end consumption, including attractive restaurants, cafes, artisanal coffee houses and retail outlets (Figure 5.10), with the apex of the Shard just visible from street level: a far cry indeed from the more diverse profile of Bermondsey Street in its

FIGURE 5.10 Artisanal consumption within the aesthetic corridor of Bermondsey Street

incipient stage of arts-inspired regeneration at the turn of the twentieth century, and with the Ticino Bakery and its gruff warnings not to impede delivery trucks but a memory.

As coda to this storyline of aestheticised redevelopment, the Ticino Bakery building was sold for £4.5 million in 2017 with its fate effectively sealed with a judgment by the local council heritage officer that the Ticino Bakery building was 'not one of the stand-out buildings' within the Bermondsey Street Conservation Area, and 'does not have any particular heritage significance'—a judgment opposed by the Bermondsey Village Action Group (situated at 14 Crucifix Lane; and reported in *London SE1 Community Website*: Bankside Press Limited: 2017).

In its place developers planned a new project incorporating a restaurant and bar and (just) seven rooms in a boutique hotel on the Ticino site: the latest in a serial experience of upgrading and aesthetic makeovers in what was one of London's most destitute neighbourhoods.

Conclusion: Culture and narratives of upgrading in London

London is often characterised as a utilitarian capitalist city replete with prosaic landscapes and the crass social attitudes associated with a dominant banking class, counterposed against the famously high-design historic cities of the Continent. But on my reading London is revealed as a metropolis with a remarkably rich cosmopolitan culture, incorporating a mix of elegance and high-design values along with signifiers of social class, ethnicity and landscape design ethos. And 'culture' is formed, produced and re-produced by means of a complex interweaving of matrices incorporating social, ethnographic, institutional, and aesthetic factors and values in the spaces of the metropolis.

Design values embodied in the landscapes and built environments of London's districts and neighbourhoods offer markers of social identity and affiliation, as well as signalling the preferences of the property owning class and the commercial sector. Up to the 1980s, local planning agencies imposed with some success design guidelines at the district and local area scale, in recognition of historic motifs and traditions.

The epitome of this control was expressed within the commercial and banking landscapes of the City; but was also manifest in the privileged residential neighbourhood communities of west-central London. But increasingly 'culture' infuses larger local regeneration programs as element of redevelopment as well as opportunity for self-actualisation.

What I've described in this chapter is a cultural representation of London that includes structuralist features, notably in the durable spaces of government, the state, religious observation and in vestigial forms at least society and class. Cultural capital embodied in social history, the built environment, education and training, creative practices and artisanal skills of populations within London's diverse districts and sites was effectively mobilised in the decades either

118 Culture and the remaking of place and territory

side of the turn of the last century in the shaping of a particularly vibrant cultural economy.

The experience of London in this century suggests that 'culture' is an embedded feature of the city, a defining expression of formal structuralism observed in many districts; but also, in some respects, a condition, a state of being, a moment in the life of the city and its constituent spaces. The spaces of exhibition and formalism clustered in Central London outwardly suggest a durability amounting to permanence, although even these are the subjects of remaking in terms of new extensions, internal reconfigurations and innovation in exhibitions and performance.

But at the local level, encompassing districts, streetscapes and sites, the resonant cultures of places like Clerkenwell, Shoreditch and Bermondey are increasingly mobilised by property concerns as critical stimuli to revalorisation: trenchant reminders of the ubiquitous potency of capital in the global city.

6

THE CONVIVIAL CITY

Spaces of spectacle, encounter and experience

Introduction: The production of convivial spaces and territories in London

London's urban structure is configured by its space-economy, segmented residential morphology and the clustering of important institutions. Government and state agencies in Westminster, and the key financial clusters of the City and Canary Wharf, exemplify the projection of power and influence in space, as described in Chapter 3, derived in part from physical form and scale as well as executive control functions.

Megaprojects incorporating cultural signifiers derived from historic buildings and layered social histories represent structural elements of contemporary London. London's mainline train stations represent both key interchange sites for the comprehensive systems of circulation in the metropolis as well as highly resonant physical complexes which contribute to the capital's stock of physical infrastructure and design cultures.

But the city also encompasses spaces of flows and circulation. These include notably quotidian flows of commuters, with the largest volumes represented in the journeys to work in the early morning and then again in the evening at the close of business. These comprise essential elements of the fabric of circulation in the city, a subject of study in its own right, as well as of public policy priority in the metropolis in regard to the financing of travel infrastructure.

Individual journeys conducted by residents of the city, and visitors of many kinds, form essential features of the urban experience. Each of these is represented in portrayals of urban life, in the present and over history. These themes were realised in L. S. Lowry's distinctive portrayals of city life in northern English cities, including paintings titled 'Coming Home from the Mill' (1928), and 'Going to Work' (1943): themes of an exhibition at Tate Britain in Pimlico titled 'Lowry and the Painting of Modern Life', 26 June–30 October 2013.

DOI: 10.4324/9781315312491-6

120 The convivial city

Society, cultures and convivial practices

The city performs as site of social interaction, including the conduct of business and political deliberations, but also incorporating convivial relations. This social aspect was prominent in the great urban centres of culture, exemplified by Paris, Barcelona and Vienna, among many other examples, but is now a feature of cities which value conviviality as community practice as well as an element of cultural policy and tourism programming.

Business districts, including Manhattan in New York, and the City of London, encompass distinctive patterns of consumption conducted within spaces of convivial interaction in cafes, bars and private clubs, involving commercial transactions as well as social exchange (Ocejo 2010). Increasingly work is conducted and performed within the 'third spaces' of cafés and bars, and indeed many of these establishments provide purpose-designed space to accommodate business professionals, technology specialists and cultural sector workers.

With the growth of tourism in the postwar era came a corresponding expansion of cultural industries, institutions, sites and spaces. And then in this century an economy of convivial interaction emerged to complement the fixed institutional sites of culture in cities. These included temporary markets and food spaces, the conversion of wholesale markets to sites of consumption and a proliferation of informal sites of cultural performance. The growth of tertiary education in cities has shaped particular forms of cultural experience, including the intermingling of youth cultures and distinctive social practices.

There are though constraints upon convivial encounter in the city in our time, not least the global pandemic of Covid-19, as well as repressive measures undertaken by authoritarian states to limit public assembly as method of maintaining social control and political hegemony.

Convivial urbanism: Situating London within urban networks and discourses

London represents one of the most expressive examples of the rise of convivial spaces and territories in the twenty-first century. Each of the established and emerging forms of conviviality are well represented in Britain's capital, and merit a place in an account of the taxonomies of space, place and territory.

Following this introductory statement, I offer a concise description of the evolving positionality of conviviality as spatial practice in the city. This discussion includes commentary on theoretical conjecture which aspires to place consumption closer to the centre of urban life, and to the economy of the city: a challenge as some would see it to a discourse which insists on the primacy of production in its multiple forms, including finance, intermediate services, cultural goods and high-value manufacturing.

I then identify what I believe are illustrative features of the social dimensions of human activity in London, including a representative taxonomy of spaces, places and territories. These include in the London case the array of high-profile sports events conducted in the city's remarkable stock of major stadia, notably for football but also for rugby, tennis and other sports, as well as annual events attracting large audiences. London also abounds in public markets and retail precincts which attract many visitors, and which perform as sites of convivial interaction.

There are of course myriad public spaces—streets, squares and memorials—which at some level signify cultural value for many and attract aggregations of users. I examine features of a selection of London's distinctive museums and galleries—each of which contributes in multiple ways to the conviviality of space in the city, but which also may present aspects of privilege and exclusion.

I also offer selective accounts of the expansion of consumption across an instructive sample of London's districts. These include high-value consumption businesses situated within zones of creative innovation and cultural production. At one level this tendency demonstrates the functional complementarity of amenity and production in zones of innovation, but also the market power (and social positioning) of consumption within the metropolis.

This co-presence of production and consumption typifies the social environments valued by workers with marked tastes and sufficient disposable incomes, as well as representing an expression of the convivial city articulated in aspirational city policies for culture and creative spaces. But as in other domains of the public sphere there are both explicit and more discrete barriers to participation, including price points as well as social coding.

Cities as sites of consumption and conviviality

Cities and towns have performed throughout history a range of specialised political, economic and ecclesiastical functions: as sites of control, administration, production and distribution. Gathering places included institutions of religious observation, markets and socialisation. Public spaces and markets in cities such as Athens, Byzantium and Hanoi afforded opportunity for social interaction and experience. The Olympic games of classical Athens provide an early example of athletic endeavour as public experience, while the ritual slaughter associated with the Roman Colosseum exemplifies another aspect of spectacle.

Over the nineteenth century, certain quarters of cities such as Paris, Vienna and Venice emerged as sites of conviviality, shaped by innovations in culture and the arts, festivals and café society. These sites and spaces afforded opportunity to share in the exchange of ideas, opinions and beliefs, as well as quotidian experiences of greeting and conversation. In these cases cultural expression and exchange, facilitated by shared consumption, shaped the identity of places and territories within cities.

There was a distinctive class and gender construction to this experience. Within the upper echelons of male society, individuals enjoyed privileges of club membership associated with income, affinity and belief structures. More open

122 The convivial city

to participation, but with entry still shaped by incomes, were the opera houses and music halls that provided venues for cultural experience. Rising incomes of the business-owning class provided opportunities to experience conviviality, in restaurants, coffee houses and, increasingly, sporting venues, as described in Thorstein Veblen's critique of the privileged leisure class (1899).

The emergence of industrial cities over the nineteenth and early-twentieth centuries among western societies, and the associated growth of industrial workers and the working class (or proletariat) is linked in the popular imagination, in novel form exemplified by Charles Dickens and Emile Zola, and in the influential polemical treatises of Karl Marx and Friedrich Engels. The life of populations situated in the working class areas of cities such as Manchester, Paris, Essen and Pittsburgh was rife with hardship, struggle and prejudice, although offering for many families better incomes than those eking out a living in impoverished rural areas.

But within industrial cities social encounter and an element of conviviality was possible, exemplified by working men's clubs and associations, although this was mostly denied to women immersed in the drudgery of quotidian housework and family care. There were though music halls, religious observance and the rise of working class sports—notably association football— to offer a measure of conviviality. For communities of faith, public ritual and assembly was organised around diocesan churches, while affinity with sporting teams, especially football, was characterised by a fierce tribal loyalty.

By the middle of the nineteenth century the arena for public experiences of spectacle had enlarged to encompass international expositions and festivals, with the London Exposition of 1851 convened in Hyde Park an early (and spectacularly successful) example, presaging a sequence of successor events carried down to the present day in ever-expanding scale. The revival of the ancient Greek Olympic Games was appropriately staged in Athens in 1896, inspired by the advocacy of Baron Coubertin.

While many of these 'hallmark events' were staged in Europe and North America, an expansion to the states and societies of Asia and South America included the Tokyo (1964) and Mexico City (1968) Summer Olympics. In this century the FIFA World Cup was staged in South Africa in 2010, representing another benchmark both in the sporting and cultural realms, and with the Rugby World Championships in Japan in 2019 a further example of the globalisation of sport.

The convivial city: Discourses and debates

The growth of the intermediate services sector, and the rise of a 'new middle class' of professionals and managers, is associated not only with a new occupational order within advanced cities, but also with a complex set of social and cultural signifiers.

These include a marked preference of this new middle class for residential forms and locations offering particular cultural values, amenity and opportunities, and a corresponding model of city planning which includes enhancements of the public realm and amenity provision (Ley 1996): a storyline with deep resonance for London (Hamnett 2016; Hamnett and Whitelegg 2007).

Other scholars point to the growth of experience and spectacle as part and parcel of the city's growing role as 'entertainment machine', coined by Terry Nichols Clark and Richard Lloyd (2004), in the wake of postindustrialism, the structural attenuation of manufacturing industries and labour, and a commensurate shift in the competitive advantage of cities from sites of industrial production to places of social experience and lifestyle.

Edward Glaeser, Jeff Kolko and Albert Saiz (2001) rank among researchers who have made a case for consumption in its various forms as core economic functions and aspect of competitive advantage, noting that high-amenity cities have generally experienced faster growth than those lacking these quality of life factors.

This view has been critiqued by scholars affiliated with the industrial urbanism school, including Allen Scott (2008), who acknowledges the increasing centrality of culture to urban development, but points to productivity differentials between industries situated with the production and consumption sectors and, relatedly, the generally low-wage levels of the latter.

It is also the case that faster-growing cities generate greater surplus value through sales, earnings and tax revenues derived from intermediate services and high-value manufacturing, generating income streams which enable investment in amenity and spending on entertainment of all kinds.

But as the twenty-first century unfolds consumption has inserted itself more insistently within the diverse spaces of the city, as cases such as Paris, New York, Toronto and assuredly London attest. Cultural quarters replete with institutions dedicated to performance have sustained their status within the spaces of the city (Evans 2003, 2009). Lively new sites and spaces of cultural consumption have emerged, shaped by public demand and supported by government investment as well as by private endowments, although the Covid-19 pandemic of 2020 has forcefully demonstrated the costs as well as benefits of globalisation.

The 'new urban ethnography' of the last quarter-century incorporates storylines of creative industries with work practices animated by convivial consumption, as Richard Lloyd (2006) recounted in the case study of Chicago's Wicker Park, and as Michael Indergaard (2004) has described in his saga of property, advertising and media companies situated in what he terms 'Silicon Alley' in mid-Manhattan, while Graeme Evans has written about the emergence of contemporary cultural quarters as product of a shift from industrial to postindustrial production (Evans 2004).

Endowments of social capital shape the nature, intensity and specific experiences of convivial encounter. Immigrants, refugees and classes of family reunification contribute to the quality of social experience in the city. Theoretical reference points include Leonie Sandercock's pathbreaking work on *Toward Cosmopolis* (1998); Martin Peter Smith's (2001) concept of 'transnational urbanism'

124 The convivial city

and the strength of diasporic-community relations; Steven Vertovec and Robin Cohen's essay on 'conceiving cosmopolitanism'; and Timothy Shortell's (2014) concept of 'everyday globalization' as evidenced by immigrant communities which present behavioural signifiers of global cultures at a lower and more informal register.

Expressions of the convivial city: Contested spaces and identities

Visitors contribute to the conviviality of the city and its diverse spaces, although they don't have the same stake in the struggles over rights to the city as residential communities and neighbourhoods. In some well-known cases the remaking of spaces for convivial consumption and experience acts to displace pre-existing residents and social groups, with Times Square (New York) and Soho (London) as important exemplars.

The spatiality of the city is entirely central to the quality of convivial interaction. There are the famous spaces and sites of cultural experience, exemplified by the Latin Quarter in Paris, South Park and the Mission in San Francisco and Siena's Piazza del Campo. These spaces enable the complex interplay of social groups and individuals in the city, including the mix of populations and visitors which contribute to diversity, while posing in some cases ethical questions about rights to the city between social groups endowed with sharply divergent resources.

To illustrate, inflows of higher-income visitors seeking consumption experiences in low-income areas of the city facilitate what some critics have termed 'poverty tourism', as observed in Vancouver's Downtown Eastside (Newman 2014), a gentrifying community which increasingly abounds with upscale restaurants. Places like el Raval in Barcelona include both low-income immigrants, as well as features of space, amenity and the built environment which attract tourists, comprising an intrusion into a socially sensitive space of the city.

London represents one of the most instructive cases for the study of convivial space, owing to its remarkable range of sites of experience and encounter, notably for culture, education and sport; to the socio-cultural diversity of community and neighbourhood, shaped by recurrent population inflows and class relayering; and to the prominence accorded conviviality in marketing strategies, regeneration programs and place-making (and remaking).

Markets—from wholesale to retail: Experiential places and territories

Many of Europe's cities and towns encompass residual features of extended histories as market towns, reflecting variously urban scale, purchasing power of elites as well as the quotidian needs of the general populace and encompassing as well 'central place' functions embedded in urban theory and history. Over time specialty wholesale markets emerged, notably meat and livestock markets, fish markets, vegetable and flower markets.

London's history as market town reflects these widely observed features of market towns and cities, but with distinctive features associated with urban scale, primacy, trade and more specially port and riverside warehousing. Specialised markets included Smithfield (meat), Billingsgate (fish) and Covent Garden (vegetables and fruit). As in other cities, the evolution and growing importance of these markets shaped space *in situ* but also territorial identities and patterns of circulation.

Over the last decades of the twentieth century new modes of marketing and distribution undercut the utility of London's wholesale markets, leading conversion to public markets of retail sales, consumption and experience, with Covent Garden the pioneer project, and now a principal attraction for many tourists and other visitors. Further, the repurposing of London's wholesale markets has stimulated new consumption and experience in proximate territories, and a modal shift in favour of pedestrian activity and (increasingly) bicycle travel.

The specific storylines of London's markets differ. Spitalfields represents a case in point, with high-margin redevelopment pressures a threat to the viability of the market in its retail form, as elucidated famously in Jane M. Jacobs' *Edge of Empire* (1996). While the area has become gentrified, and while upscale shops have located on adjacent Brushfield Street, Spitalfields retains its vibrancy as a place of shopping, informal dining and social experience.

Portobello in the Notting Hill District of Kensington & Chelsea represents an exemplar of a successful street market experience which has evolved but maintains an identity both as site of consumption and experience. Portobello has its origins in the nineteenth century as a food market, but by the mid-twentieth century transitioned to specialise in clothing and antiques, and has evolved as England's largest antiques market. The Saturday market continues to attract large volumes of visitors (Figure 6.1) including both a mix of local residents, visitors and tourists.

The Borough Market located in Southwark represents a particularly vivid storyline, owing to its history, situation and trajectory. The market has its origins 1,000 years ago, as a site of trade in foodstuffs and other commodities near the only bridge across the Thames to the City of London and was comprised of unlicensed trading, and with entertainment including bear-baiting.

The Market was reopened in 1756 as a fruit and vegetable wholesale market. A viaduct of the London and South East railway cutting though the Market in 1862 represented a substantial disruption, although at the same time increasing accessibility for some populations.

The Borough Market experienced a rebirth in this century, shaped by specialty retail and growing consumer interest in artisanal foods, associated with London's growing new middle class and its specific consumption tastes and preferences. And over the last decade or so the provision of fresh English food products has been accompanied by high-value products from diverse regions of France, Italy

FIGURE 6.1 Portobello Saturday Market, Notting Hill

and Spain, presenting an international flavour to complement domestic offerings, and underscoring the internationalisation of consumption and experience in the globalising metropolis.

The Borough Market, proximate to Southwark Cathedral and the London Bridge Underground Station, and lying in the shadow of the Shard, occupies a place of prominence, and has performed an important role in the overall regeneration of this district of South London. The upscale nature of food stalls and product lines also underscores the dramatic class change in South London, as Bermondsey was one of the poorest districts in London for much of its history. The recent revival of the Bermondsey Antique Market (Figure 6.2) offers another marker of the role of convivial places in the area.

In common with other prominent London locations, the Borough Market has also served occasionally as locus of violence, with the terrorist attack of June 2017 that claimed eight lives as the most exigent example, and demonstrating that no part of the capital is immune from the disaffections and radicalism of the age. The Borough Market wasn't specifically targeted as the locus of violence against civilian populations. But the enactment of death within the spaces of the Market, and the mixed local and international roster of victims, serves to underscore the possibility of violent encounter as well as convivial experience in public spaces within the global city.

FIGURE 6.2 The Bermondsey Antique Market, Southwark

Convivial experiences and iconic institutions

The growth of major cultural institutions, galleries, museums and exhibition space, over the eighteenth and nineteenth century in major cities, represents an important space-shaping and territorial form of development. These structures were designed initially to safely house important collections of art and artefacts of many kinds, and to provide spaces of specialised preservation and study by experts. From the mid-nineteenth century onward, these institutions emerged as sites of public spectacle and experience. Major institutions such as the Louvre in Paris, the Uffizi in Florence and the British Museum in London contributed powerfully to the urban cultural imaginary as well as the field of spectacle.

Many of these important institutions are implicated in the strategic reshaping of space, place and territory in the space, owing to a mix of factors that can include scale, design and grandiose ornamentation in situ. They can be associated with the development of ancillary uses, such as souvenir shops, artists supply stores and cafes, and thus contributing to the larger reterritorialisation of urban space. These include the Palazzo Pitti in Florence's Oltrarno district; the Quai d'Orsay in Paris; and the Hermitage in St Petersburg.

In London the National Gallery occupies a prominent place in Trafalgar Square—itself both an iconic and increasingly contested space—while the

128 The convivial city

British Museum is the institutional centrepiece of Bloomsbury. And as I described in Chapter 5, the density of cultural institutions in Camden form an archipelago of exhibition and related conviviality for many visitors.

Among the principal cultural institutions in London is the Victoria and Albert Museum, situated on Exhibition Road. The V & A is the world's largest institution dedicated to design and the arts. But its significance as space-shaping influence dates to its prominence in the development of 'Albertopolis', in the remaking of this precinct of South Kensington as a residual benefit of the extraordinarily successful 1851 Great Exhibition of the Arts and Industry. As is well-known the V & A and its companion institutions the Museum of Natural History and the Museum of Science were funded in part from the income generated by the 1851 Great Exhibition.

The V & A encompasses galleries of important design products and creations, arrayed by geographical provenance as well as societies and cultures of origin; admission contributions on the part of visitors are encouraged but not required, opening up the exhibitions to all regardless of income. The V & A also includes appealing products for sale in its shop, always attracting a crowd of casual visitors as well as buyers, contributing to the conviviality of space and possibilities of human interaction within the gallery.

The V & A also includes attractive consumption spaces within its building envelope as well as within the open areas of the principal quadrangle. Each offers a distinctive experience of conviviality. When the weather is cooperative large crowds congregate within the seating areas of the quadrangle, enjoying edible treats at reasonable price points (Figure 6.3), while the interior dining spaces include the cafeteria and the beautiful Morris Room, with pianists adding to the cultural ambience of this sublimely cosmopolitan experience.

There are though signs of more problematic features. While admission to the V & A is open to all, unobtrusive observation of visitors suggests a dominant middle-class quality among the galleries' patrons. While this class configuration is by no means outwardly oppressive, there can be questions as to how comfortable low-income visitors might feel in the presence of conspicuous consumption and affluence.

The new Exhibition Road Quarter and entrance, designed by Amanda Levete Architects, with most of the £55 million cost supported by four principal donors, offers an attractive and inviting option for visitors, in contrast to the more formalist Victorian elegance of the Cromwell Road entrance. The new entrance also aligns the V & A with the original motif of a spatially (and visually) integrated vision of 'Albertopolis' in South Kensington financed in large part with the surplus revenues of the Great Exposition of 1851, articulated by Henry Cole among others.

But just outside the V & A, special security forces clad in protective gear and armed with submachine guns patrol Exhibition Road—a reminder of the potential of violence in the public spaces of the metropolis.

FIGURE 6.3 Cultural experience and convivial consumption at the Victoria & Albert Museum

Symbolisms of sites—contested memories and postcolonial meaning

London is perhaps more than most global cities replete with places of celebration and commemoration, attracting in many cases large flows of tourists and other visitors. As the capital of a unitary state engaged in protracted wars and contestations, many of these sites commemorate struggle, including, notably, victory in war and signifying battles, including Britain's experience of the two world wars of the twentieth century.

Paris, Rome and perhaps Washington DC among western states and societies are the only comparators to London in the extent of commemorative places and sites signifying a public record of conflict and struggle. Visitors, both domestic and international, are invited to give witness to these commemorative spaces, places and institutions. Major thoroughfares, the riverbank and other places and territories within London encompass a range of statues, plaques and other diverse commemorations of historical struggle and conflict.

Waterloo Station commemorates the definitive battle (1815) of the Napoleonic wars and is deemed objectionable by some French cultural arbiters. And Berlin may be unique among European cities in its commitment to recognising the sufferings associated with the darkest moments of the twentieth century, and more especially the human costs of totalitarian visions for which Berlin served as political command centre.

130 The convivial city

While there is certainly no shortage of places of commemoration of earlier conflicts and struggles, including many sites associated with the extended conflicts with France from the seventeenth century onward, there is an exceptional extent of commemoration of Britain's struggles in the two world wars of the first half of the twentieth century, augmenting the continuing filmic and literary treatments of these epochal conflicts.

Over the current century there has been at least a measure of historical redress, or at least recognition of the instructive value of social struggle amid the greater extent of more triumphalist mementos. There are also memorials to the unique evil of the Holocaust displayed in London, including a deeply moving bronze tableau ('Kindertransport') of five Jewish children selected for transport to Britain prewar, situated in the Liverpool Street Station forecourt.

Trafalgar Square represents an example of a place which combines for many visitors an evocative site for conviviality within its capacious public spaces, in the form of a convenient place for meeting. Trafalgar Square also offers opportunity for cultural expression, in the inscription of temporary (chalk) art on the horizontal spaces and both scheduled and impromptu musical performances. The Landseer lions project the power of imperial Britain, dominating the overall landscape of the square.

But (literal) prominence in Trafalgar Square is given to the famous Nelson's column, with Britain's most celebrated admiral perched 52 meters above the Square. Here the master narrative has taken the form of Admiral Horatio Nelson's victories over the French and Spanish navies, most notably at Trafalgar (21 October 1805), which ended the threat of invasion.

But writer Afua Hirsch contributed a thoughtful piece on the harsher meanings and symbolism of Nelson's column in the Guardian online (22 August 2017), titled: 'Toppling Statues? Here's why Nelson's Column should be next', referencing both the Royal Navy's implication in colonialism and the slave trade, and evidence of Nelson's implicit support for prominent slavers. Hirsch's essay generated over a 1,000 email replies—testimony to the emotive power and very mixed symbolism represented by a famous place of commemoration in the centre of the former British empire.

Football—changing registers of affinity and territorial identity

London represents among other registers of global status one of the world's principal venues for sports, both professional and amateur, comprising in the aggregate a principal arena of convivial association for many. At the apogee of sporting enterprise London has hosted the Olympic Summer Games three times: 1908, 1948 and 2012. For the 2012 Olympics, important investments in new infrastructure included the Olympic Stadium at Stratford, near the eastern terminus of the Jubilee Line in East London, converted to the London Stadium, West Ham United's home field.

In terms of major sports, professional and amateur, London is home to the national Rugby stadium at Twickenham in southwest London, hosting the world championship in 2015; as well as two iconic international cricket grounds, Lords and the Oval. The Wimbledon tennis championships in South London are the last of the four elite international tennis tournaments to be played on a grass surface.

London also hosts a major international running marathon every summer, while the Oxford-Cambridge Boat Race, although essentially a private contest between England's two oldest universities, attracts crowds Thameside in the hundreds of thousands every April. Dozens of competitive rowing clubs and boathouses line the Thames, both in East London, and also in West London from Putney up to the Teddington Lock: the upstream limit of the Tideway.

These examples represent just a few prime markers of London's status as international sporting centre: to be sure an important measure of the convivial city of the twenty-first century, and of course these are augmented many times over in terms of the vast numbers of participants among London's universities, colleges and schools, as well as the deep pool of community based sporting activities throughout the territories of the capital. But (association) football arguably occupies for many pride of place among spectator sports in London.

Although for many cognoscenti of association football Manchester and Liverpool occupy the pinnacle of English football culture, with Liverpool FC and Manchester United boasting exceptional records of success at the highest level of national and European competition, and while cities of the Midlands (notably Aston Villa and Wolverhampton) and northeast (particularly Newcastle United) ranking among successful clubs, London is entirely unique in its density of professional football clubs not just in England but internationally.

As for other British cities, there is a distinctive territorial aspect to the location of football stadia and (to an extent at least) the concentrations of club members and supporters, as well as the cultural inflections of supporters in each case. Socially football was considered a working class (and working men's—few women habitually attended professional sports) pastime in London, in contrast to rugby (associated with public schools) and cricket, and indeed admissions to (mostly standing room) stadiums was in relative terms affordable.

Famous clubs were located in East London (West Ham United, Millwall and Leyton Orient), and North London (Tottenham Hotspur and Arsenal, with the latter relocating from Woolwich in southeast London to Highbury in North London). Fulham was located riverside in southwest London, but the area was increasingly working class at the time of club formation (1879).

Registers of change: Class, social affinity and property development

The recasting of football in London as business enterprise and social practice has been driven by a number of broadly linked processes. The establishment of the English Premier League (EPL), replacing the old Football League First

132 The convivial city

Division and attracting enormous payments to clubs for television rights, has greatly privileged the more successful clubs in the League, with London clubs such as Arsenal, Chelsea and Tottenham attracting a global following.

The locus of the 'engaged' supporters (or fan) base for the most prominent clubs has effectively been rescaled to include contingents within North America, Asia and Africa: representing a diaspora of club followers, and a market for team merchandise. The big London clubs (and other prominent EPL teams) embark on global tours of exhibition matches in the (ever-shorter) summer off-season, in part to acknowledge their offshore fans, to 'build the brand', and not incidentally to generate new revenue streams.

Second, the fundamental class restructuring of London and more specifically the mass incursion since the 1980s of middle-class residents within former industrial districts and communities in East London and north London especially has transformed the territorial base of Club affinity. These clubs still pay homage to their working-class origins in official histories, but increasingly cater to the professional and managerial class cohorts who comprise their season ticket-holders.

Empty seats at some home games attest to the diminished loyalties of a portion of these more affluent fans, perhaps reflecting a less committed relationship between clubs and ticket-holders, as well as the greater lifestyle and recreational options for contemporary football ticket-holders relative to the deeper commitments and club affinity of working-class fans of the last century. At the same time, certain pubs in London cater to football audiences by televising live matches, in effect offering a convivial option for fans seeking a communal (and affordable) viewing opportunity (Figure 6.4), as social experience and corporate branding exercise.

Refreshments on offer within London's principal football stadiums are more diverse—and pitched at higher price points—than the limited fare of previous eras. Ironically, perhaps, Arsenal's former emblem (in use, 1949–2002) at a time when its support included a large working-class population included a detailed shield and cannon, and a Latin insignia, 'Victoria Concordia Crescit' ('Victory grows out of harmony'), while the current badge comprises a simplified cannon motif and the word 'Arsenal'—presumably more legible to an international audience.

Third, the capital relayering and insistent upgrading reaching ever-further into areas of London beyond the Central Area has decisively shifted the urban land economics underlying profitability and returns to capital, stimulating new football stadium construction with proximate ensembles of high-value residential, retail and consumption uses.

In consequence a number of London teams and other EPL clubs have embarked upon programs of new stadium construction or, in other cases, comprehensive upgrades to facilities in situ. In important cases this entails a redefining territorial shift, as in West Ham United's move from Upton Park in the heart of East London to the (former) Olympic Stadium at Stratford, while in others—such as

The convivial city **133**

FIGURE 6.4 Football mad, football crazy

Arsenal's relocation from Highbury to Holloway—the distance is more incremental, but involves a re-imaging of stadium configuration and optics to suit a more affluent market base.

In the case of West Ham United, the London Stadium (formerly the 2012 Olympic Stadium) provides a far-larger capacity than its predecessor, Upton Park, but lacks the intimacy of the latter, where the field of play extended almost

134 The convivial city

up to the stands, and where the fans of the 'Hammers' enjoyed a unique level of engagement with the team on the field.

'Territory' is still in use as a marker of location for London teams, evidenced in an imaginative display at Lillywhites (Piccadilly Circus) in 2018 which showed the location of each London club on a grid displaying specific degrees of longitude and latitude (Figure 6.5).

The construction of a new stadium for the national football team in North London offers some instructive insights into the role of stadia in comprehensive redevelopment programs. The new Wembley is located next to the SSE arena, built for the 1934 Empire Games, last used for international sporting events for the 1948 London Olympics, and now a listed Grade II building deployed for exhibitions and other private events (Figure 6.6).

But the new Wembley, and more particularly investments in transportation and local amenity, has attracted capital for development, including residential buildings for Middlesex University. And the new Wembley is host not only to home matches for the English national football team, but is also home for regularly scheduled games of the (US) National Football League, demonstrating the global reach of London as base of sports (Figure 6.7), while the London Stadium has played host to games between the Boston Red Sox and New York Yankees of the American League.

FIGURE 6.5 Situating the Tottenham Hotspur Stadium, showing geospatial coordinates

Source: Lillywhites Piccadilly Circus

The convivial city   **135**

FIGURE 6.6   Residuals of the sporting past: The SSE arena, Brent

FIGURE 6.7   Display of NFL football scarves, New Wembley Stadium concourse

Millwall FC: Territory, class and redevelopment narratives

Millwall FC, situated in London SE16, operates within the second tier of English football, the Championship. Its name recalls its origins in the Isle of Dogs in Docklands, although the club left the site over a 100 years ago, and now plays in southern Bermondsey.

Millwall earned a fearsome reputation as a club with followers endowed with a predilection for violence, but it now appeals to local families in the area, including a special dedicated stand in the ground, the 'Lion's Den' (Figure 6.8). In the twentieth century Millwall contested with some success against the giants of English football, but now principally operates as a feeder team for larger and more affluent clubs in England.

The marketing of Millwall FC incorporates some gestures to the past as well as a forward-looking connections to the community. One of the numerous Millwall scarves available for purchase at the Den's shop includes the message 'No-one likes us'—an ironic gesture to the Club's rougher past.

Inevitably perhaps the tentacles of gentrification have extended to the larger area which encompasses the Millwall grounds and training site, as upgrading tendencies seep further south in Southwark, and encroaches upon the neighbouring borough of Lewisham. In discussion with a Millwall FC employee I was advised that families are now the preferred target demographic (conversation 25 September 2018).

FIGURE 6.8 The Den, Bermondsey, London SE16

There is still older council housing in the area, but there also is a substantial residential redevelopment trajectory. A local school, Ilderton Primary School, situated a short distance from the Millwall Den, features signage attesting to its commitment to excellence in teaching, and perhaps reflecting a larger community social aspiration, with the simple message that 'we are outstanding' as marker of aspiration.

The profitability of land resources in southeast London generated proposals by Lewisham Council to acquire Millwall's grounds, with a view to high-value residential development. This saga continued over an extended period from 2012, and included consideration of a compulsory purchase order which would have evicted Millwall FC. But in a council decision 10 October 2019 Lewisham Council elected not to proceed with the purchase order, thus affirming the tenure of Millwall FC in its current grounds.

Neighbourhood festivals: Community, social capital and sustainability

Ideas of community structure have been influenced in part by the Chicago School of social ecology and other morphological models, where neighbourhoods are shaped by labour markets, occupation, race and ethnicity. While variants can always be acknowledged, and while no district can be impermeable to change, the underlying assumption is that planners and scholars can gain an understanding of the composite spatial features of the city through a careful study of how labour and work reproduce a class system which finds spatial expression in the city, influencing zoning and land use planning choices and decisions.

In turn, social relations at the neighbourhood level reflect the class and ethnicity based values and behaviours of majority groups, with households and the family as cornerstone concepts and underpinning social relations and network formation at the community level. At the same time markets and festivals can offer opportunity for a more diverse array of cultural groups and actors, and for a lively expression of the cosmopolitan character of globalising cities and societies.

London abounds with local markets and food stalls which offer both a measure of cultural expression and business opportunity (Figure 6.9), and opportunity for convivial experience.

In some cases the scale and structural organisation of markets encompass both food stalls as well as local crafts and artisans, which represent a 'destination site', as in the case of the Camden Lock Market (Figure 6.10) which occupies a compact site at the northern end of the Camden 'cultural archipelago', discussed in the previous chapter. London's markets also afford opportunity for collaboration which contributes to the formation of social capital at the community scale.

In Britain, sociological studies have included both work relations and family dynamics as key factors in shaping community and neighbourhoods, as evidenced in Willmott and Young's classic study of Family and Kinship in East London (1957). In working class communities men forged connections with

FIGURE 6.9 Food stall, Spitalfields district

FIGURE 6.10 The Camden Lock Market

fellow workers within factory systems, and with local pubs and football teams as key institutions of collective ritual or behaviour, while women householders participated in informal neighbourhood relations and performed leadership and linking relations with kin: in the aggregate comprising elements of the social capital of the community.

Postindustrialism, gentrification and immigration rank among the principal factors that have fundamentally reshaped the social geography of the city since the 1970s, especially within the large tracts of working class territory characteristic of advanced urban societies. While features of the built environment, certain institutions and social memory offer a measure of continuity, the restructuring of the city has in many cases subverted the traditional dynamics of community interaction.

In response cities have promoted institutional investment and innovation as a means rebuilding social capital, including community centres, playgrounds and event programming, with each offering opportunity for constructive interaction. Local festivals have emerged as one of the most encouraging of these concepts. While there is of course great variation in terms of themes, periodicity and scale, local festivals typically offer an opportunity for bringing together both new and more established cohorts within the community.

In London the rich mix of local populations shaped by occupation, class, ethnicity, gender and belief systems among other variables has supported a proliferation of local festivals and related community events across the metropolis, animating the spaces, places and territories of the community.

In addition to the enjoyment of local festivals 'in the moment' festival events can contribute to community building, in an era of often rapid change and household turnover. In this regard Nancy Stevenson has contributed to a deeper understanding of the purpose and meaning of local festivals in East London, by examining two instructive cases shaped in part by official regeneration policies associated with the legacy programs of the London 2012 Olympic Games. Stevenson's broader remit takes the form of an inquiry in part designed to assess whether local festivals in two East London sites linked by the Queen Elizabeth Olympic Park, Hackney Wick and Fish Island, have succeeded in building social capital and sustainability capacity in two neighbourhoods experiencing pressures of change.

Stevenson's research on these East London sites is shaped by two critical research questions: first, inquiring into the type of social capital developed in the host community engaged in festivals; and, second, identifying implications of social capital for host communities in 'emerging destinations'. She establishes that prior to the 2012 London Olympics the area under study included substantial areas of social housing and industrial estates, and an incipient creative cluster. But this creative element was largely subsumed, and the 'conviviality' associated with conventional cultural quarters was lacking.

Nancy Stevenson provides an account of two festivals imbued with different organisational motivations and operating characteristics. First, the Wick Festival

140 The convivial city

incorporated a participatory model designed to bring in community members broadly. This clear purpose led to the awarding of substantial grants, including a £1 million award from the 'Big Lottery Fund', deployed in Hackney to provide local groups with grants, social investments, loans and microfinance. A second group, 'Hackney Wicked', operates a festival convened over a weekend in August, and includes studio openings, films, tours, theatre and music.

Stevenson's field research disclosed that 'both festivals have a role in developing social capital' (Stevenson 2016: 8). Further, both festivals are 'multifaceted' and designed to appeal to a broad range of people on a number of levels, helping to build a sense of place. But a concern arises from her study associated with the unequal distribution of benefits, as she finds that the 'accrual of social capital exacerbates existing inequalities within the host community' (Stevenson 2016: 990) which favours the 'haves' over the 'have nots' in this instructive case.

Nancy Stevenson's insights have value in contributing to forming an analytical lens for assessing the deeper social value of festivals linked to mega-events such as the London 2012 Olympic Games. In the following chapter I explore (then Prime Minister) David Cameron's efforts to link the development of London's innovation economy situated in Shoreditch to the Olympic Park.

Conclusion: Conviviality in a time of pandemic and political crises

I have elected in this chapter to identify 'conviviality' as spatial practice as well as an expressive form of social interaction in London. As a framing device for discussion I selectively developed critical perspectives on six portfolios of convivial interaction which I believe serve to illustrate important arenas of socio-cultural contestation as well as convivial experience in contemporary London: streetscapes, public markets, cultural institutions, symbolic sites, football stadiums and neighbourhood festivals.

London abounds in a remarkably rich array of spaces and sites for convivial activity, some enabled principally by business and private actors and institutions, while many others are fostered in various ways by the state, including a prominent place for convivial opportunity in local regeneration policies and cultural programs.

As for other global cities such as Paris, Rio de Janeiro and New York among many others, the cultural richness and diversity of social groups is effectively mobilised for tourism and, increasingly, for competitive advantage in attracting new investment and business formation. In this regard 'conviviality' is part and parcel of the repertoire of aspirational programs for cities seeking investment as well as the growth of cultural tourism.

The significance of convivial practices in space encompasses but extends beyond the enjoyment of co-mingling of social groups and individuals, to include aspects of exclusion, contestation and displacement. At the localised scale public markets aspire to sustain a diversity of representative displays reflecting the rich mix of communities and populations.

But increasing competition for (necessarily) limited exhibition space may contribute to a general upgrading in place over time. And the fortunes of high street shopping, a mainstay of social life in London as for other British cities, are increasingly fraught, with online retail sales gaining ground across many product lines, and with Covid-19 exacting costs which include many outright firm closures as well as sharply diminished sales volumes for others.

For some traditional forms of convivial practice and association the upgrading trajectory has been very steep indeed. This is especially the case for London's principal football clubs, which have experienced a spectacular increase in admission prices owing to the globalisation of professional sports in general and the market success of the EPL in particular. At a lower register of scale within London's professional football landscape Millwall's experience has been instructive: the past is selectively acknowledged in publicity pieces, but the aspirational tendency includes a larger place for families as feature of community marketing.

And Millwall's security of tenure in the New Den was for a time thought to be at risk owing to the appetite for land resources for high-margin development associated both with revenue-seeking local authorities, as well as market players: demonstrating the ubiquity of the property market throughout London.

7

SPACES OF INNOVATION

The 'Tech City' trope and place branding in London

Introduction: The urban-innovation nexus in history

Cities have performed as critical sites of innovation throughout history, across a spectrum of fields including science, technology, medicine, governance and the arts. Innovation as deployed by business privileges change which drives efficiency (and higher margins) in the market. But 'innovation' in the larger sense, including the fields of health, education and social services, enables experimentation and discovery which generate novel processes and systems: ideally connoting higher social value, and enhancing human welfare and quality of life.

Cities have deployed advantages of specialisation, agglomeration, social capital and key institutions to foster innovation throughout key periods in history. These qualities comprise collectively the platform (or milieu) for innovation. But crucial factors in the fostering of innovation also include individuals of talent and genius; the stimulus and exchange value of cultural diversity; and civic leadership.

The saliency of innovation to the economic welfare of cities in this century has stimulated a new era of city place-marketing, incorporating inducements as means of attracting propulsive firms, investment capital and talent: the latter including entrepreneurs and artists.

To illustrate the positionality of cities within important innovation epochs, Athens was host to a remarkable society of philosophers, civic leaders and scientists, underpinning the formation of the Hellenic society and epistemic community. Florence in the *quattrocento* was the European epicentre of innovation in arts and science, shaped by a unique concentration of artists of genius and supported by the city's great wealth based on banking and finance, comprising a system of patronage in support of cultural and scientific innovation.

Hanoi functioned as the centre of an extended regional arts and crafts region over extended history, as depicted in the scholarship of Sarah Turner (2006),

DOI: 10.4324/9781315312491-7

while Kyoto has served as a site of excellence in the synergy of arts and technology in craft production for centuries (Goto 2012). Other examples of the positionality of cities at their moments of peak influence across diverse fields of innovation include Berlin as 'electropolis' after 1840, Vienna for innovation in music, the theatre and psychiatry 1780–1910 and Manchester as the first advanced manufacturing city 1760–1830: each case elucidated by Peter Hall in his influential compendium *Cities in Civilisation* (1998).

In this chapter I set out some important aspects of London's trajectory as centre of innovation, with particular reference to the ways in which innovation plays out within space. Aspects of novelty involve new enterprise formation, emergent divisions of labour and an increasing deployment of digital technologies in production processes and market penetration.

What we observe in London's innovation experience in many cases takes the form of a recombinant form of development that synthesises creativity and cultural assets; technology and human agency; and resonant imageries of place and urban history, supported by aspirational policy framing.

I present an exploration of some key features of innovation as development trajectory in London, emphasising the interdependency of actors, institutions, place and territory. First I sketch the contours of London's saliency as site of innovation, emphasising the importance of space and territory. Examples include the pioneering role of South Kensington, the milieu effects of medical research institutions in Camden and the symbolism of Google's European headquarters at King's Cross: a 'landscraper' extending parallel to the station's railway platforms, and representing in scalar terms a far cry indeed from the intensely localised debates and narratives around the first stage of redevelopment in this iconic London district, described in Chapter 4.

I offer a narrative of twenty-first-century technology sites in inner north-east London, largely imposed within creative industry districts which had earlier displaced informal communities of artists who represented the initial agents of gentrification.

For this discussion I draw on my field research in London as well as scholarship which offers insight on critical aspects of the digital economy in north-east London: Jo Foord's work on 'risky experimentation' conducted among diverse firms and actors clustered near the Silicon Roundabout; Julia Martins' insightful research on the networks of places and environments which form the critical milieu for the intense networking and social relations which underpin the digital economy; and original work by Max Nathan, Emma Vandore and Georgina Voss on the sometimes fraught relations between the state, policy and place-branding exercises in the Tech City initiative.

These lines of scholarship represent individual research interests, and, collectively, instructive contributions to a critical discourse on the city as site of innovation. In each case scholarship is informed by literatures on the industrial district, contemporary urban ethnography and relatedly the social workings of innovative firms and finally the scope (and limitations) of the territorial configuration of the digital economy in place.

144 Spaces of innovation

A conclusion suggests the outlines of London's evolution as contemporary centre of innovation, replete with upgrading and dislocation effects as well as experiences of discovery, invention and productivity enhancement.

The resurgence of cities as innovation centres

In the postwar era technological innovation was largely situated within extended regional territories, associated with multiple sites and leading research universities. At the forefront of these regional innovation clusters were two leading American exemplars: Silicon Valley in northern California, centred on Stanford University; and the 'Route 128' innovation district in Massachusetts, with MIT a centrepiece: with some common operating characteristics but also signifying differences in research cultures (Saxenian 1991).

Within Europe, a leading innovation cluster was centred in Cambridge University, while the Paris-Ile de France region functioned as a territory of advanced innovation in France (Storper and Salais 1997). In leading Asian economies, notably Japan, industrial innovation was conducted in exurban universities and science parks, such as Tsukuba Science City situated in the Ibaraki Prefecture.

Regional innovation districts continue to perform across a spectrum of fields including science, technology, communications, aerospace, medicine and applied economics. But the late-twentieth century encompassed an expansion of science-based innovation within cities. The digitalisation of information content enables the conduct of research and development within compact premises in the city, exploiting advantages of skilled labour, agglomeration economies and proximity to universities, while at the same time sustaining contact with geographically distant collaborators and markets.

The fusion of culture and technology in production processes, allied with both the concrete features and semiotic values of place, are key features of innovation sites within the metropolis, as Trevor Barnes and I have shown in the Seattle case (Barnes and Hutton 2016). While the larger Seattle-Central Puget Sound Region includes Microsoft, Boeing, Amazon and Starbucks, there is a complex ecosystem of innovation shaped by districts such as Fremont, Ballard, Belltown and South Lake Union, each of which accommodates a range of enterprise from start-ups and small and medium-sized enterprises (SMEs) to global corporations.

Legacy effects of postindustrialism in the spaces of the city

The concentration of older, labour-intensive production sectors in cities such as Chicago, New York, Montreal and Manchester was by the 1970s a hallmark of industrial obsolescence, disinvestment and competitive disadvantage, and cumulatively as harbingers of overall decline. London was particularly disadvantaged by its industrial structure, exacerbated by disinvestment and chronically poor management-labour relations.

The collapse of traditional industry in London exacted devastating social costs upon manufacturing labour and families, and a legacy of structural unemployment. But other residuals of postindustrialism in London included compact spaces of disinvestment, and (mostly) brick warehouse buildings and former craft production spaces, comprising a distinctive built environment suitable for adaptive re-use.

Cultural legacies of London's industrial past included stories, social memory, polemical treatises and a rich fictional literature situated in iconic districts and communities. These have been selectively redeployed in place-marketing by property concerns, and by the sequence of arts, cultural and digital technology firms which have colonised these spaces.

Immigration was a key factor in the development of craft production regimes within London's City Fringe and inner north-east districts in the nineteenth century, as it is in the rise of the British capital's contemporary urban innovation economies. New entrants arrive equipped with skills critical to the deepening of human capital and the sharing of specialist knowledge.

There is diffusion of innovation between centres of high-level innovation and knowledge production, exemplified by epistemic community linkages between cities such as San Francisco, Seattle, New York, Hamburg, Kyoto, Mumbai and Singapore. Each of these cities is markedly multicultural—a key to attracting 'talent' and aspirational young people embodying cultural diversity, and receptivity to new experience and opportunity.

Recasting space, place and territory for the innovation economy

The first two decades of the twenty-first century witnessed the transformation of older industrial areas in cities of the Atlantic realm. The storyline takes the form of new high-density residential precincts, together with complementary investment in amenity and the quality of the public realm, as well as evocations of the 'new industrial district'.

As Stefan Krätke has observed in his essay on 'New economies, New spaces' (2015), an urban development trajectory comprising in varying measures the arts, culture, technology, talent and generational change has underpinned the revival of high-value production industries in the city. In this regard the collapse of traditional industry over the 1970s and 1980s, while producing massive social costs including structural unemployment, and representing an essential precursor to rounds of gentrification, has also opened up many of the spaces of the urban core and inner city to new rounds of investment, innovation and regeneration.

The location of innovative agencies and enterprise in cities is shaped in part by the local state and by incentivised locational programs. The competitive dimension of high-value innovation programming was illustrated in the intense competition for the second location of Amazon, headquartered in Seattle's South Lake Union district, with New York the eventual winning bid. A second contest

146 Spaces of innovation

appeared in the offering as the deep divisions in New York over the costs of hosting Amazon has led to the latter revoking its decision, on the grounds that New York is not sufficiently needy to appreciate the opportunity.

The frantic scrambling for positioning among cities large and small for the Amazon prospectus surely demonstrates the illusory nature of place-positioning for many marginal candidate cities aspiring for a place within the global roster of tech-cities. Intense competition for the second Amazon city site presents a counterpoint to the dramatic falling away of Olympic candidate cities in an era of civic austerity, with Milan-Cortina-d'Ampezzo declared the winner from a small field of contenders for the XXV Winter Olympics.

The summer Olympics is still viewed as a hallmark event for aspirational states, cities and governments, owing in part to legacy value of post-Olympic sites in the city, relative to the more dispersed winter games sites, and of course the larger event scale and media opportunity of the summer games.

The MaRS research and development cluster in downtown Toronto offers a prominent example of a strategic-scale, technology-led project situated within the central area of a major metropolis. Established in 2008, the MaRS complex incorporates four linked buildings in the heart of the city, bounded by College Street and University Avenue. MaRS is situated in proximity to the University of Toronto's St George Campus, major specialist and teaching hospitals, and local and provincial government offices. Toronto's MaRS is a not-for-profit project, offering applicants a range of services for start-ups and early-growth enterprises situated within four clusters; health, financial technology (fintech), and clean tech and enterprise, supported with $3.2 CAD billion in start-up funds.

The twenty-first-century innovation economy: London as exemplar

London, which represented one of the worst cases of industrial disinvestment and collapse in the last century, has emerged as one of the most important staging grounds for a narrative of regeneration shaped by intersections of the arts, culture and technology. The instrumental redeployment of the resonant spaces of the inner city for high-value innovation represents a strategic element of development policy in the British capital.

London's 2012 Summer Olympic Games is broadly viewed as successful in terms of visitations and revenues (and the high performance of many British athletes), although critics properly drew attention to the opportunity costs of public spending on spectacle in a city which had already benefitted more than other British cities from Government infrastructure investment. But the plans for regenerating the former Olympic territories in East London for job generation and community benefit are by no means fully realised, despite major investment *in situ*, and attempts to link the Park to the innovation economy situated in Shoreditch.

London has emerged in the twenty-first century as a leading European innovation centre across a range of specialised fields. Importantly, space, place and territory are implicated in the formation and operation of innovation in London, reflecting clustering tendencies for agencies seeking collaborative synergy, networking opportunity and the efficient provision of specialised services.

Within London there are major research universities, institutes and teaching hospitals which host important innovation programs and projects, as well as centres of music, drama and other performative arts. There is a particularly salient cluster of health and medical research institutions situated in Camden and the City of Westminster, including agencies associated with University College London, and national-level specialist hospitals and treatment centres, as well as the Royal Society of Medicine and the Royal College of Surgeons (Figure 7.1)

Technology districts clustered within London's City Fringe and larger inner city territory represent fields of inter-sectoral experimentation and inter-firm collaboration. For many of these institutions and programs, partnerships with other actors and institutions within Europe has been a crucial stimulus to collaborative enterprise—now jeopardised by Brexit and the prospective revocation of key programs which have facilitated cooperative institutional relations and inter-firm partnerships across diverse sectors.

London performs as metropolitan centre of a global-scale extended innovation region situated within southeast England, incorporating leading universities as well as numerous medical research centres, technological institutes and business incubation centres. London, Oxford and Cambridge comprise the apex sites of a 'golden triangle' of innovation across key sectors and fields, with dense channels of collaboration between specialists situated in each place

In this regard Oxford has traditionally been seen as specialising in humanities and social sciences, but has emerged as a leading university for medical research, and excels in physics and mathematics, while Cambridge has developed global excellence in engineering and applied science, built on a platform of its stellar basic science reputation. London leads in fintech ('Financial technology'), and has deep strength in engineering (particularly situated within the Imperial College—South Kensington district) and medicine, supported by the teaching hospitals and many medical research institutes.

Innovation in London: Transportation and the reshaping of place, territory and milieu

London's identity has been formed in large part by the political power shaped by an exceptional degree of primacy in a unitary state, by the control and exploitation associated with its role as capital of empire and by the specialised banking and financial institutions which have underpinned London's status as apex-level global city.

London has also performed as an important centre of innovation across a range of fields of endeavour, which have worked to reshape space, place and territory

148 Spaces of innovation

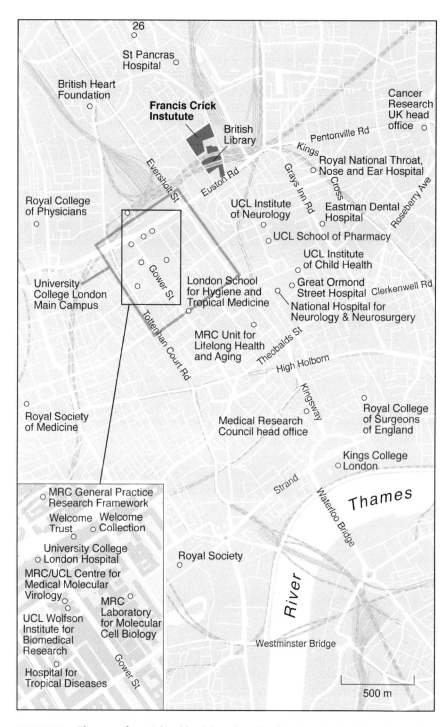

FIGURE 7.1 Clusters of specialised health and medical institutions in Central London

in the metropolis. Innovation in transportation systems and infrastructure has been instrumental in reconfiguring space and travel behaviour. These include notably the world-leading metropolitan Underground system (1863), which shaped space by (effectively) shrinking distance by reducing travel tine within the London region, while facilitating the growth of commuter communities on the Fringe of the London County Council area and within the Home Counties.

More recently, the early introduction of central area congestion charges in London (2003) stimulated more sustainable transportation modes, emphasising as policy preference public transportation systems over private automobile use, and including the expansion of a system of bicycle lanes, consistent with larger social values and a political commitment to the green city agenda.

The Crossrail high-speed system extending east-west across London below the Underground system represents another important example of systems innovation, as well as a massive capital expenditure, in the region of £15 billion— criticised by many outside the London-southeast region as another instance of the metropolitan bias at Westminster. Crossrail (to be branded as the Elizabeth Line once operational) comprises an important development phase of London's Underground system, but also features significant novelty as well, and thus comprising another exemplar of innovation in this 2,000 year old settlement and city.

The scope and complexity of the engineering for Crossrail represented certainly one feature of innovation, as did the commitment to preserving the remarkable archaeological assets associated with the project. In this regard approximately 200 archaeologists were engaged in a project to record some 10,000 objects spanning the early settlement record of London, disclosing new evidence of Londinium and its pre-Roman antecedents. This unique experience of archaeological discovery and preservation have been addressed in a series of publications which comprise key cultural and historic aspects of what has been the largest European urban infrastructure project of the twenty-first century.

The institutional platform for London's innovation sector: Government and non-state actors

Key institutions and linked agencies within specific districts comprise important territorial platforms for innovation within London. These include notably major research-based universities, such as University College London, King's College, the London School of Economics and Political Science, and importantly the Imperial College of Science, Technology and Medicine, which anchors the South Kensington culture and innovation district, along with the Victoria & Albert Museum. These major universities, established during the seminal rounds of investment in higher education in the British capital in the nineteenth century, represent principal sites of innovation in science, medicine, technology, the arts and humanities.

In the last half-century, many more universities and colleges have been established in the capital: these have extensive commitments to teaching and learning,

150 Spaces of innovation

but most also possess the capacity to engage in research and development—adding to the institutional diversity of London as a centre of knowledge-creation and innovation.

A feature of innovation in the present century takes the form of a stronger institutional commitment to fusing culture, design and engineering. South Kensington has performed as critical staging ground for the promotional synergy of science, the arts and industry, observed in the 1851 Exhibition in Hyde Park, and the subsequent deployment of accumulated revenues to the development of key institutions along Exhibition Road. More recently Imperial College has collaborated with the Royal College of Art to offer an Innovative Design Engineering (IDE) program.

A recent demonstration of this institutional commitment to innovation at the intersection of science and the arts is seen in the establishment in 2014 of the Dyson School of Design Engineering at Imperial College. Design Engineering is Imperial's tenth engineering department, promoting innovation in research, teaching and industry liaison, proclaimed in the College's promotional materials as 'the fusion of design thinking and engineering knowledge and practice within a culture of innovation and enterprise' (http://www.imperial.ac.uk/design-engineeeing/; accessed 19 February 2019)—another expression of the 'cultural turn' in advanced innovation experiences situated within the spaces of the city (Figure 7.2).

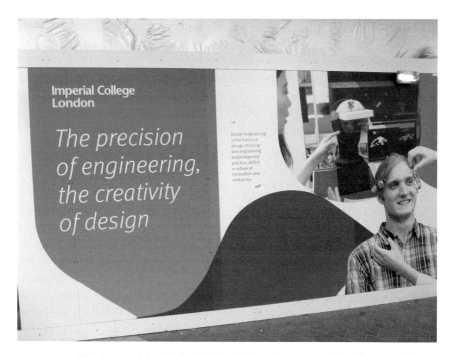

FIGURE 7.2 The Dyson School of Design Engineering, Imperial College London

Government in London has become increasingly engaged in the promotion of the capital as site of innovation, and more particularly research in telecommunications, life sciences and clean energy technologies. The innovation sector has been a feature of policy promotion in London, at the level of the Greater London Authority (GLA) and many of the London boroughs, as well as finding expression in local regeneration programs.

The GLA's commitment to fostering London's role as global centre of innovation was exemplified in in the establishment of a London Office of Technology & Innovation (June 2017), with a view to facilitating stronger collaboration among players in this sector, and a companion project in the form of a London Office of Data Analytics.

As the information release (17 September 2018) acknowledges, London cannot claim any particularly pioneering role in this institutional launch, as it was inspired by the 'first mover' example of New York under Michael Bloomberg's mayoralty, and indeed Bloomberg Associates 'advised London on the UK capital's new strategy' (London Innovation: 'London to launch new office of innovation'; story by Jonathan Andrews 17 September 2018. http://www.london-innovation. org.uk/; accessed 19 February 2019).

At the local level, innovation is promoted by larger boroughs such as Westminster, Camden and Islington, and is increasing encouraged in local regeneration programs. The Borough of Camden, for example, has supported innovation within its area in part through a project labelled 'Steam'—encompassing science, technology, engineering, arts and mathematics, and prioritising opportunities for Camden's youth to gain critical skills and opportunities in the innovation sector broadly.

Much of London's innovation economy is concentrated in boroughs north of the River, although there are to be sure vibrant cultural districts and residential communities in South London (Chapters 5 and 6), as well as high-impact megaprojects which leverage cultural memory for market advantage (Chapter 4). Thus a proposal for a South London Innovation Corridor represents an aspirational vision for a consortium of boroughs including Lambeth, Lewisham, Southwark and Wandsworth, with a view to deploying the collective endowment of talent, institutions, the built environment and quality of space to attract innovative investment and enterprise.

The strategic centrepieces of this program proposal (submitted to the Mayor of London's 'Strategic Investment Pot') entail investment for two central clusters (situated at South Bank and Vauxhall Nine Elms) and designated 'local growth clusters' (Brixton, New Cross, Old Kent Road, Peckham, Camberwell and Wandsworth). The project information piece ('South London Innovation Corridor', Lambeth Council https://www.lambeth.gov.uk/better-fairer-lambeth/project/south-london-innovation-corridor; accessed 20 February 2019) notes that the partnership bid also 'brings together universities including Goldsmiths, London South Bank University, the Royal College of Arts, and

152 Spaces of innovation

Capital Enterprise – London's membership body for connectors, investors and policy makers supporting London's start-up scene'.

Spaces of innovation in London: Processes and exemplars

There are aspects of innovation within London's business complexes, expressed in the development of new financial instruments and business practices, and observed notably within the City of London and the global business and finance-scape of Canary Wharf. These include the packaging of property development and investment, and the marketing of these high-margin instruments in domestic and foreign markets.

I acknowledge though Stefan Krätke's (2014) astute judgment that much of this notional innovation and creativity takes the form of mimicry of established practices within the 'dealer class' which forms a very substantial component of the commercial sector in capitalist societies. There is also experimentation and innovation at lower registers within the broadly-defined business sector, including novel forms of product display and marketing as the decline of traditional high street retail continues.

There are crucial 'milieu' effects at play in London's spaces of innovation, within which the formation of institutional platforms of talent is conducive to experimentation and innovation. As I demonstrate later in the chapter, place, territory and talent are important in forming diverse ecosystems of innovation. Key institutions accommodate many fields of experimentation; these include universities, specialist research institutions and teaching hospitals, which contribute both to innovation and to place-making in the metropolis.

Allied to these are London's many museums, exhibition and performance spaces which contribute to place-identity, but which also offer possibilities of innovation across artistic and professional realms. Exemplars of relations between place, actors and innovation include the distinctive cultural milieu of Bloomsbury in the first half of the twentieth century, and the film-making industries in Soho and Ealing.

Larger clusters (or in some cases corridors) of innovation in the scientific and cultural domains, taking in exhibition of specific applications, include South Kensington, with Imperial College and the Victoria and Albert Museum as prominent institutional centrepieces (Figure 7.3); and a cultural corridor in Camden extending from the British Library southwards toward the British Museum, and including the Tavistock Institute as well as a range of other agencies, discussed in Chapter 5.

There are spaces of innovation which comprise a mix of studios, workplaces and 'third spaces' of temporary or informal innovative practices. Cafes and coffee shops offer space for extended visits by company workers and freelancers as elements of their marketing and business identity—and thereby diversifying the milieu of collaborative innovation in the city.

FIGURE 7.3 Display of innovation in specialised wood applications, Victoria & Albert Museum

London's historical wholesale markets, repurposed to sites of the arts, exhibition and consumption, include opportunities for cultural experimentation, notably in music, crafts and visual arts, as well as meeting space for those engaged in dialogues and conversations concerning opportunities for creative enterprise.

Spaces of innovation in London: Specialisation, splintering and succession

Territorial patterns of innovation in London reflect larger tendencies in the space-economy of the metropolis in the digital era. There are signifiers of the traditional industrial district and its refinements over the postwar era, including complexes of specialised production and divisions of labour, buildings and infrastructures and information systems which facilitate sourcing of key inputs, as described by Robert Lewis in his depiction of industrial districts and factory networking in Chicago (Lewis 2008).

An important tendency takes the form of the impact of innovation on the construction of space in the metropolis, and in particular the impact of advances in telecommunications. There is a literature which emphasises the potential of technological innovation to destabilise, reorder and 'splinter' space in the metropolis, elucidated by Stephen Graham and Simon Marvin (2001) two decades ago.

154 Spaces of innovation

Graham and Marvin were concerned about the 'splintering' effect of advanced technology systems on the variegated social spaces of the city. But an important element of the storyline comprised functional interdependency: acknowledging 'that much of the "urban" is infrastructure; that most infrastructure actually constitutes the very physical and sociotechnical fabric of cities; and that cities and infrastructure are seamlessly coproduced, and co-evolve, together within contemporary society' (Graham and Marvin 2001: 179).

Andrew Harris (2013) has written about the deep equity issues around access to urban infrastructure within highly polarised global cities, including London and Mumbai: constituting a form of socio-economic inequality as well as splintering along social and more particularly class lines.

Two decades following *Splintering Urbanism*, it seems clear that the diffusion of advanced telecommunications can facilitate the concentrating effect of digital clusters in global cities, including London, as well as global-scale innovation centres such as San Francisco, Boston and New York. Digital capacity enables spatial diffusion of many forms of innovation, and the production and dissemination of knowledge across global space. But advanced telecommunications can also enable compact sites of innovative enterprise to sustain complex input-output relations with distant co-producers and markets.

Advanced communications systems, including the Internet, the Cloud and emerging forms of data collection, storage and transmission, have profoundly reshaped the geography of enterprise and innovation across space, and by extension the loci of innovation. But what remains as constant is the capacity of space at the district level to provide both a physical infrastructure for enterprise over extended cycles of change, as well as a productive milieu conducive to collaborative experimentation.

Narratives of technology-based 'new industrial districts' in London

Innovation represents a potent force for industrial succession and dislocation in the metropolis. The cultural signifiers of community history, local amenity and the built environment attract technology-intensive institutions and enterprise in London. These form the materiel for a distinctive public policy storyline in the London case, as well as an exemplar of industrial succession *in situ*.

Inner north-east London represents a significant terrain for the study of industrial restructuring, with particular value for understanding processes of change in the metropolis over multiple innovation cycles. Broadly, the narrative of change within north-east London is shaped by new enterprise formation, novel products, emergent divisions of specialised labour and place-branding. There is also continuity including, notably, innovative practices that draw on pre-existing assets and imaginaries of place.

To elaborate: what has characterised the mix of industries over successive restructuring experiences within inner north-east districts of London since the

1840s is: (1) a predominance of small to medium-size enterprises, mostly in the production, distribution and communications sectors; (2) divisions of labour comprising craft workers, including younger workers and apprentices as well as smaller contingents of managers and professionals; (3) a marked spatial density of production, housed mostly in two- to five-storey buildings suitable for adaptive re-use; (4) the temporal 'extensification' of work in response to the exigencies of demand and tightly contracted work orders; (5) local consumption features, which together with the mix of industries and labour, and places of religious observation and other institutions, form part of the identity of the area; and, finally (6) pressures of upgrading on firms and labour derived from changing product cycles.

Continuing with this longer view of the systems of innovation and structures of production within inner north-east London, I now turn to signifying aspects of contemporary change, drawn from field study in the present century. These include: (1) innovation in production and communications technologies that characterise the twenty-first-century urban economy; (2) the infusion of talent, skills and knowledge contributed by immigrants and visiting students within the district's labour force, contributing to a productive diversity of human capital, cultures and work skills; (3) the co-location of an extensive base of consumption amid networks of innovation, including sites patronised by local workers both for leisure and collaborative work sessions; (4) the elasticity (or porosity) of district borders as processes of upgrading produce territorial capture and re-imaging by property market agencies; and, (5) the role of government and private agencies in place-remaking, promotion and marketing, as sites transitioning from artisanal craft production to digital industries represent important features of regeneration programs.

Innovation as trajectory: Upgrading, place-remaking and marketing

The storyline of the inner north-east of London, from the arts-based regeneration of the 1980s, to the contemporary re-imaging of place and industrial structure projecting an insistent technological identity, tracks in some important ways experiences of other iconic sites, perhaps most notably the recasting of the South of Market Area in San Francisco.

Since the 1980s the South of the Market Area (SOMA) has transitioned from an established working-class territorial configuration; to an increasing presence of artists and galleries, restructured by the innovations and dislocations of the dot. com era in the 1990s; and through to the contemporary technology-intensive profile which meshes with the larger narrative of the Bay Area as the world's leading region of technological innovation (Hutton 2008).

London and San Francisco also share a common thread of change expressed in political values: from an earlier concern for public welfare, social equality and opportunity, and the virtues of a mixed economy, to state commitment to

156 Spaces of innovation

technology as process, product and place-imagery in the pursuit of competitive advantage. There is also a corollary shift in local planning, from regulation in the interests of sustaining a balance of land use, and protecting the rights of marginal groups to space in the city, to programs for promoting higher-margin activities.

In this interpretation, 'space' as a resource for capital relayering and redevelopment, aligned with trajectories of globalisation, supplants the more complex (and contested) notion of 'place' in the city, where social rights, community values and residential tenure are recognised and to a degree at least are accommodated in local policy regimes and planning regulations (as elucidated in the San Francisco case by Solnit and Schwartzenberg 2000).

The instructional value of inner north-east London as case study of innovation also lies in its thematic intersectionality, traced along a span of restructuring episodes which includes both effects of global-scales processes, and the contingency of specific sites, described in Chapters 5 and 6.

As London's innovation economy expands within the inner north-east zone, the initial territories of arts and digital experimentation retain their descriptive power and explanatory value. The rise of Shoreditch as critical space of cultural innovation over the 1990s, with its epicentre in Hoxton Square, incorporates features of novelty as well as more widely-observed phenomena. The lineaments of the district include a distinctive street pattern and built environment comprised of former tailoring, furniture and related industries and labour dating from London's heyday as national centre of light industry and warehousing, depicted in Figure 7.4.

Figure 7.4 incorporates defining features derived from a program of site visits, interviews and cultural mapping I conducted in the area, including initial visits and scoping in 2000 and 2002, and intensive field survey work 2003–2006. A record of field observations enabled the construction of an activity profile for the area, as well as a baseline for assessing change in the area 15 years on from this initial field work exercise.

I identified three specialised ensembles of activity nested within the precincts of the area, together with intricate production networks, consumption amenity and social activity. The Hoxton Square precinct incorporated concentrations of artists and designers; galleries, notably the White Cube; and presented images of lively social interaction.

Second, the 'Shoreditch Triangle' (Figure 7.4) featured a stronger business orientation along Curtain Road and Great Eastern Street, reflecting its proximity to the City of London, with businesses including telecoms, training institutes and property concerns, as well as upscale restaurants, bars and pubs.

But the internal street patterns within the Triangle accommodated a more intimate array of spaces and users, including galleries, shops, cafes, bars and pubs—comprising an exceptionally lively creative habitus for artists, students and visitors (Figure 7.5). The business profile of this sub-area included graphic designers, artists' studios, interior decorators, new media and arts-oriented retail operation.

Spaces of innovation 157

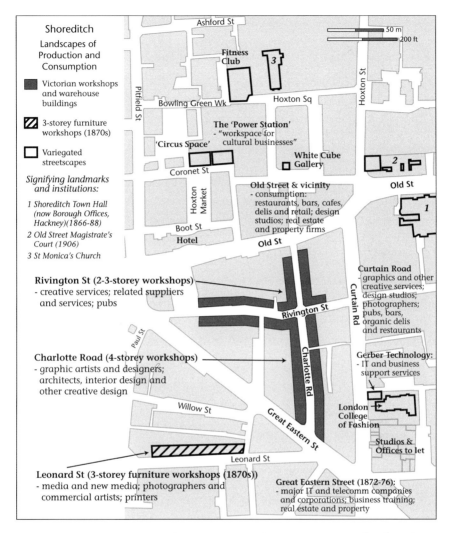

FIGURE 7.4 Hoxton and the Shoreditch Triangle, London Borough of Hackney circa 2004

Source: Hutton 2008

A third structural element of the area included in my survey took the form of furniture warehouses along Leonard Street (Figure 7.6) dating from the 1870s (Cherry and Pevsner 1998: 525) which accommodated design firms (architects, interior design, graphic artists and designers) as well as technical business support enterprise such as commercial photographers.

Hoxton and the larger Shoreditch district occupy an instructive position within the annals of urban change in the twenty-first century, associated with a rich layering (and relayering) of the arts, artisanal production and digital enterprise

158 Spaces of innovation

FIGURE 7.5 Convivial spaces of the Shoreditch Triangle

FIGURE 7.6 Leonard Street, Shoreditch, leading to Great Eastern Street

and labour. Part of its value for urban studies has to do with its signposting of change, in the form of recurrent upgrading and dislocation, and reflecting at ground level the larger benchmarks of restructuring in a global city.

Each phase of development in Shoreditch over the past four decades, in the form of the arts, culture, applied design and cultural production and (presently) digital industries, accompanied by a distinctive consumption culture, has been displaced by (and to an extent incorporated into) higher-margin industries and labour. I now draw on the work of scholars who deploy innovative research methodologies to interrogate the critical lineaments of change in this key district.

The Tech City project centred initially on the Old Street Roundabout comprises an instructive case in the potential and limits of innovation-based regeneration. From the initial arts-inspired cultural regeneration process in Shoreditch in the 1980s, the Tech City initiative embraced an extensive set of institutions, with varying degrees of commitment.

The participants in this venture included corporations such as Google, Facebook, Intel, Microsoft and Cisco, as well as major London universities, notably Imperial College, Queen Mary University and University College London. Participating local governments included the London Boroughs of Hackney, Newham and Islington. What follows is a discussion of the lessons of the Tech City (and larger regeneration aspirations) of innovation in Shoreditch, drawn from the work of influential scholars and comprising an important discourse of contemporary urbanisation in a globalising context.

Jo Foord on the transition from 'Creative City' to 'Tech City'

As Jo Foord has shown in her insightful essay on 'The new boomtown? Creative city to Tech City in east London' (*Cities* 33, 2013: 51–60), the outcomes of the project to rebrand Shoreditch as site of technology-driven innovation are shown to be quite mixed. Foord depicts a complex suite of factors which include notably the residuals of the global financial crash of 2008. Effects included opening up opportunities for the entry of new players within the strategic spaces of the metropolis—including tech-startups.

As Foord recounts, a harder business model and profit-seeking edge to the enterprise structure of global cities in the aftermath of 2008 shaped a transition from the original arts-based regeneration of the City Fringe and postindustrial inner city, to a more revenue-driven 'business-to-business' (b2b) operating model for the actors within the cultural sector. Citing Andy Pratt (2009), Foord acknowledges a transition in political attitudes toward 'culture', from an ensemble to be valued for its intrinsic self-actualising potential and community-building benefits, to the instrumental value of creativity in economic development and competitive advantage.

Jo Foord's storyline of industrial succession is informed by analysis of 261 firms clustered near the Silicon Roundabout (Figure 7.7). She offers a critique of the idealised model of technology clusters as semi-autonomous agents of

160 Spaces of innovation

FIGURE 7.7 The 'Silicon Roundabout': Intersection of Hackney and St Luke's, Islington

innovation and growth, arguing instead that the vitality of Tech City 'emerges from risky experimentation across co-located sectors in which hitherto unrelated knowledges and activities (for example, software and advertising) are being combined' (p. 52).

Foord acknowledges some of the conceptual, definitional and measurement problems associated with assessing the implications of both official and survey-based data, including the variegated hybridity of firm types: '[t]he blending of digital technology and creative development within firms makes it increasingly difficult to distinguish tech-firms from content companies' (57)—a recurring problem of scholarship on inner city innovation districts since the onset of the so-called dot.coms for almost three decades, and articulated in Richard Lloyd's (2006) sobering account of the limited novelty of 'new media' as industrial prototype in the spaces of Chicago's Wicker Park.

Seen in the wider context of London's digital economy which comprises a wide swath of clusters, districts and firms across London, the Tech City ensemble represents in Foord's cogent analysis 'a contingent assemblage of actors and sectors: an experimental production space offering opportunity and favouring opportunism' (Foord 2013: 59). This profile of a new industrial district in London apparently in flux contrasts with the relatively robust nature of specialised production in earlier iterations of the industrial city.

Juliana Martins on the innovation economy as social construct: Evidence from Shoreditch

The complex ecosystem of innovation in Shoreditch underscores the utility of the city as site of factor sourcing and production and, relatedly, the value of amenity and consumption *in situ* as complements to enterprise formation. The literature on the saliency of inner city sites imbued with resonant cultural imaginaries for creative activity is well-established, with a distinctive conceptual framing informed both by established industrial geography and location theory, and evocative case studies (Scott 1997, 2000; Markusen and Schrock 2006; Ho 2009).

The social dimension of the innovation economy situated in resonant spaces of the city incorporates possibilities of stimulating interaction conducted within street-level consumption sites, exemplified by an array of establishments which abound in Shoreditch (Figure 7.8). There is a powerful human impulse to socialise in spaces of the city even in the most trying circumstances, as the persistence of co-mingling in the pandemic of 2020 demonstrates. But less is known about the specific and instrumental use of internal spaces of consumption in the contemporary innovation economy, and more specifically the connections between internal space and specific work practices.

FIGURE 7.8 The 'business after business' scene in the innovation economy: Goose Island Brewpub, Shoreditch High Street

162 Spaces of innovation

Juliana Martins has made an important contribution to our understanding of the more specific value of space, place and the built environment in work practices in Shoreditch, derived from careful field work and interviews.

Martins adopts a production-centred approach to understanding space and environment in the digital economy. She takes as her particular remit the need to advance the field by investigating the networks of sites where networking, innovation and co-production take place, and in articulating the 'spatial characteristics of semi-public and public places, access and control, location, and attributes of the neighbourhood' (Martins 2015: 126).

Julian Martins deploys a methodology which takes the form of semi-structured interview with digital industry workers, observation of spaces within the study area and secondary data sources in the form of maps, reports and websites. The value of her research lies in the complex structures and systems of thematic intersectionality, categorised by 'work activities' (producing, meeting, networking, learning, displaying work and socialising), and 'types of space' (workplace, residence, semi-public spaces, events and public spaces) (Martins 2015: 130).

This typology points to a detailed articulation of the distinctive space-economy of the digital age: one in which the idea of the industrial district retains value, but implies a more complex industrial and labour geography than was typical of the factory era in its heyday circa 1800–1950 in Britain, and in other economies of the North Atlantic realm.

At the same time, many (likely most) digital workers also use their homes for some part of their work, representing the spatial and temporal 'extensification' of work in the new economy, reported by Helen Jarvis and Andy Pratt (2006) in the important San Francisco example, and in field study by the author (Hutton 2008: 192–218) on the origins of the live-work and work-live landscapes and social relations in the SOMA district of San Francisco.

In Juliana Martins' typology the 'base' (company office) and ancillary spaces in proximate areas comprise the milieu of innovation and co-production in Shoreditch. The company office, as disclosed in Martins' research, retains its utility in the digital economy as a unique place where workers can both meet with clients and interact with each other.

But as Martins affirms, 'ancillary spaces' (semi-public and public places) provide critical complements to the base. For the digital workers in Shoreditch, access to ancillary space beyond the company office is associated with the specific time and purpose of work functions; the level and type of interface required; and the time/context issue. As Martins observes, the 'advantages and limitations of each space in relation to the requirements of these work tasks are critical to understand how the extended workplace operates in digital production' (Martins 2015: 131).

While literatures on both the cultural economy and emergent technology district underscore the intimate relations between work practices and consumption amenity (Hutton 2016; Ocejo 2010), a particular value of Juliana Martins' scholarship lies in the granular detail concerning affinity of workers for particular kinds of places, derived from observation and interviews. Digital economy workers in Shoreditch evince preference for 'ancillary' meeting spaces according to factors of difference including the provision of tables, chairs, lighting and design features, as well as the capacity to secure space for effective work periods and meetings with colleagues and clients.

What might appear superficially as random distributions of workers patronising the rich amenity base of Shoreditch, Juliana Martins' program of careful field work discloses as clear evidence of intent in the selection of work sites and meeting places according to purpose, implying a micro-geography of innovation and production.

Max Nathan, Emma Vandore and Georgina Voss on place-branding and spatial imaginaries

Through a succession of field research models and methodologies deployed to assess the workings of 'new industrial districts' in the creative economy and digital age, researchers have disclosed interdependency between production regimes and the programmatic deployment of place-imaginaries. The evolution of place-identity in cities is associated with local values and social practices associated with race, ethnicity and class—key elements of urban ethnography and sociology—as well as influencing the formation of industries, divisions of labour and work practices.

The apparent shortening of innovation and restructuring cycles experienced by cities under neoliberal regimes, and pressures and opportunities of globalisation, has produced among other things a corresponding relayering of territorial identities and imaginaries. Further, these territorial imaginaries are incorporated within external marketing programs by state agencies.

At the urban scale there has been a transition from cultural quarters based in part on neighbourhood cultures (Evans 2003, 2004); to a profiling industrial districts as cultural districts and sites (Hutton 2016: 235–285); and to an emphasis on the re-imaging of place derived from cultural signifiers, as Sheng Zhong has demonstrated in the Shanghai case (Zhong 2012). So there is perhaps a back-and-forth emphasis on interdependency between culture, space and technology over the last three decades, producing often contradictory, contested or incoherent place narratives.

Max Nathan, Emma Vandore and Georgina Voss (2018) take this storyline further in an incisive analysis of the progress and limits of the place-branding effort underlying the development of the East London digital economy. What distinguishes this experience from others in London includes the role played by central government and more specifically David Cameron, (then) prime minister of Britain; the resources allocated to place-branding and marketing; and the territorial over-reach of the area deployed in the Tech City development program.

Nathan, Vandore and Voss open with a summary statement of the evolution of place-branding specifically as an external marketing device, drawing on a stream of influential literature starting with Sharon Zukin's *Loft Living* (1989[1982]) and *The Cultures of Cities* (1995) and through to research published in the present

164 Spaces of innovation

century, typified by Andy Pike's 'economic geographies of brands and branding' (2013). They take as remit for their study of East London a clear need to investigate 'the mechanisms by which place brands operate in real-world policy contexts' (Nathan, Vandore and Voss 2018: 409), and extending the field to include recent examples of urban technology clusters.

As Nathan, Vandore and Voss affirm, previous studies of place-branding have focused on signifying aspects of sites, clusters and actors, and have mostly been 'sector blind' as expressed in the analytical model and field research program. Accordingly they situate their case study within the lineage of 'post-industrial clusters', incorporating an understanding of path-dependency which nonetheless allows for 'technological branching' into fields related to the initial structure of production and labour.

In this interpretation, clusters can contain 'many previous "versions" of themselves' (Nathan, Vandore and Voss 2018: 412)—rather than a 'pure' taxonomy of sectors, industries and enterprise type often positioned as the most desirable ensemble set out in development strategies. This aspect of variegation is shaped in part by the resilience of firms shaped by earlier eras of innovation, the co-location of enterprises with in some cases weak linkages and elements of co-dependency, the overlapping of policy regimes and the assortment of institutions which include property firms and estate agencies.

This complex palette of enterprise was evident in the case of South Park in San Francisco circa 1995–2005, during which period a pioneering wave of the so-called 'dot.coms' coexisted not only with a landscape of live work studios but also machine shops which had formed important elements of the Bay Area economy over the twentieth century (see Hutton 2008: 187–202; and Hartman and Carnochan 2002 for deeper historical treatments of change in SOMA).

The East London case study includes an acknowledgement by Nathan, Vandore and Voss that 'we know little about which policy mix is most effective' (Nathan, Vandore and Voss 2018: 413) in shaping the configuration of industries, firm types and labour in place, although 'branding' is increasingly popular in promoting the development of existing sites (or to generate new ones). There is also a problem in the management of clusters where both positive feedback (knowledge spillovers and labour pooling) and negative feedback (competition, crowding and cost increments) are often co-present.

As Nathan, Vandore and Voss acknowledge, the succession of policy experiments in East London have an extended lineage in the postindustrial era, from the 'City Fringe' mix of top-down metropolitan regeneration programs involving a promiscuous assortment of industries and labour; residuals from the industrial economy of London ravaged by the disinvestment and restructuring episodes of the late-twentieth century, together with an assortment of robust cultural industries; and in this century what Nathan et al describe as the 'playful Silicon Roundabout' program.

Such was the power of the Silicon Roundabout imaginary that it captured not only the interest of local regeneration professionals and borough planning staff, but also David Cameron. The British Prime Minister was much impressed

Spaces of innovation **165**

with the apparent vigour and robustness of the London tech economy clustered around the Old Street Roundabout, in contrast to Britain's slow recovery from the ruinous financial crash of 2008 and its deep effects observed in key sectors such as banking, property and commercial development.

At the launch of the East London Tech City venture in November 2010 (Figure 7.9), Cameron sketched the contours of a program comprising three

FIGURE 7.9 Signage proclaiming the advent of 'Tech City', Shoreditch

166 Spaces of innovation

objectives: building the cluster through awareness-raising and light touch business support, attracting FDI and 'connecting' the Shoreditch arts and digital technology scene to the Olympic Park after the 2012 London Summer Olympic Games. Cameron's Tech City program featured a branding exercise, refitting buildings post-Olympics and leasing these to technology multinationals and major universities: a vision reminiscent of the famous 'developmental state' storyline of Singapore, but lacking the policy tools, sustained industry-government nexus, corollary investment and sustained political commitment of the tropical city-state.

In order to tease out the outcomes of the Tech City program, Nathan, Vandore and Voss implement a research program including ethnographic observation, a series of semi-structured interviews with firm representatives, and then semi-structured interviews with members of the policy community. The storyline encompasses a rich depiction of industrial succession from the positionality of Shoreditch as an important district specialising in light manufacturing in the furniture, clothing, textiles, printing and jewellery sectors, just north of the City, and north-east of the Clerkenwell areas which specialised in precision instruments design and manufacture (Ainsworth 2010).

Following the orthogenetic redevelopment of Shoreditch as a key site of industrialisation, arts-based regeneration and cultural innovation, the Silicon Roundabout label as Nathan, Vandore and Voss see it represented a storyline of succession, shaped by an infusion of technological innovation in new enterprise formation.

The Tech City branding, with its imprimatur of central government and the Prime Minister, was seen by many as an official endorsement of the Shoreditch innovation ecosystem, although as Nathan et al acknowledge it is difficult to assess how much of the growth of the sector since 2010 is attributable to government agency.

But other outcomes included the almost inevitable capture of value increments in the area by the property sector, including financial institutions as well as estate agencies prevalent within the area. This in turn reflects the earlier experience of market capture of land price increments ensuing from the revalorising impacts of the arts and cultural imagery of (hitherto) frontier districts in inner London, as observed in the cases of Clerkenwell and Bermondsey described in Chapter 5.

As Nathan, Vandore and Voss pointedly acknowledge, the high-level aspiration of the Tech City program as elucidated by David Cameron, in the form of an ambition to extend the zone of innovation from Shoreditch to the Olympic Park in Stratford, fell well short of the policy targets. This was in large part owing to the gap between the official government vision of 'proximity' as depicted in policy mapping, on one hand, and, on the other, the on-the-ground distance between the core site of Silicon Roundabout to Stratford of approximately 30–40 minutes combining public transport and walking distance.

There are of course examples in cities such as London, New York, San Francisco and Shanghai of an apparent 'travelling' of culture-infused technology innovation sites across city space. This phenomenon is in part a response in part to the displacement effects of serial upgrading, as well as the search for place-authenticity on the part of artists and other creatives. But prominent examples of multiple innovation sites within a single city owe more to the characteristics of these sites (notably Mid-Manhattan, Chelsea and Brooklyn in the case of New York), and their capacity to generate their own 'scenes', than on agency efforts to link developmental imaginaries over extended space.

There is considerable capital flowing into Stratford. But these inflows have little to do with the state's aspirational program of incorporating Stratford within the spatial ambit of inner north-east London's technology sector, and far more to do with the profit potential of high-density residential sites and ancillary retail and consumption.

To the extent that Stratford's development trajectory is 'aligned' with larger trajectories of development within East London, it is more compellingly situated within an extended north-south property development zone encompassing Stratford, the Olympic Park, Canary Wharf and the Greenwich Peninsula, as I discussed in Chapters 3 and 4 of this volume, than it is within a largely imaginary technology corridor.

Conclusion: Place, territory and ecologies of innovation in London

London demonstrates the rise of the global metropolis as innovative city-region, encompassing multiple zones of enterprise and collaboration and shaping a complex space-economy comprising new platforms of institutions and networks.

London exemplifies innovative practices encompassed within the defining categories of space deployed in this book. These include the importance of 'place' as habitus for key actors and social cohorts, incorporating the built environment, social histories and community identity—each of which performing as prompts to creativity and innovation; and 'territory', encompassing extended urban fields of collaborative social groups, and information flows and linkages.

Territory as spatial construct for innovation is exemplified in the clusters of medical and health-related institutions in Camden, and cultural industries and creative workers in Soho and Clerkenwell, as well as the appeal of Shoreditch which resonates for artists, cultural actors and most recently digital industries. The industrial Shoreditch which featured mostly small- and medium-size firms engaged in specialised craft production, with material inputs sourced locally, has been supplanted by digital enterprises which can readily source inputs both from local suppliers and more distant ones via the Internet and the Cloud.

Advantages of metropolitan cities such as London for a broad range of innovative enterprise include universities and other knowledge-producing institutions; deep endowments of both financial and human capital; and in this case productive collaborations between London and other British and European agencies and firms. These attributes have in the aggregate comprised platforms for new

168 Spaces of innovation

enterprise, and shape a productive diversity of creative ecosystems in the heart of the global city. London's innovation sector also incorporates major operations of global corporations such as Google and Facebook

Within London, the spatiality of innovative enterprise includes the diverse platforms of universities and colleges, clusters of fundamental and applied sciences and specialised medical research centres and health care delivery systems. As a current example (September 2020) the MRC Centre for Global Infectious Disease Analysis at Imperial College is viewed as a leader in research on the 2019-nCOV virus, in collaboration with the University of Oxford, while researchers at Imperial's St Mary Campus are engaged in the development of a safe and effective vaccine against the virus.

The emergence of innovative enterprise in inner north-east London as described in the three case studies represents a serial relayering of sites initially activated by arts and culture in the closing years of the last century, then transitioning to applied design and its commercial applications. The rise of institutions and enterprise favouring digital technology based within key sites and spaces in London can be viewed as a classic demonstration of succession shaped by experimentation and risk-taking as well as mimicry of established practices.

Through the cycles of innovation, upgrading and restructuring in inner north-east London over the past century and a half, the critical importance of 'space' as effective field (or conduit) of information flows, knowledge exchange and collaboration of many kinds remains a constant, as affirmed in the three instructive case studies drawn from influential scholars described earlier.

As a closing observation I suggest that advantages of culturally-diverse metropolitan cities like London are situated not solely in the co-presence of key institutions and firms, but also in the possibilities for individuals and smaller groups to contribute to incremental change and often serendipitous progress across complex fields of human endeavour. Social capital accrues at different scales in the city and can offer possibilities for self-actualisation as well as productive collaboration. There is scope for individual human agency, informal association and temporary collaborations engaged in the search for innovation within the spaces of the city, even as the global reach and power projection of mega-corporations such as Amazon, Facebook and Microsoft expands.

8

NEW GENTRIFIERS AND EMBLEMATIC TERRITORIES

Hypergentrification, displacement in situ and territorial dislocation

New gentrifiers and emblematic spaces in London: Introduction

London was a pioneering site of gentrification in the postwar era, with social upgrading at the neighbourhood scale observed as early as the 1960s, and, along with New York, has constituted an instructive staging ground of social change into the present century. As the geographical field and spatial ambit of social upgrading continues to expand into new territories, and with scholars developing fresh storylines and explanations, London retains its saliency as exemplary site of gentrification, as pioneering model and as comparator.

In his essay on 'gentrification as global urban strategy', Neil Smith cited London as a first example of process and problematic: '[w]hereas for Glass, 1960s gentrification was a marginal oddity in the Islington housing market – a quaint urban sport of the hipper professional classes unafraid to rub shoulders with the unwashed masses – by the end of the twentieth century it had become a central goal of British urban policy' (Smith 2002: 439). Key assumptions associated with this approach to neoliberal governance include the ostensible benefits of attracting higher earners and investors as agents of international connectivity, the stimulus effects on residential development and markets for other private goods and contribution to the cosmopolitan ambience of hosting cities.

Andrew Harris (2008) has emphasised the critical need to understand contrasts as well as commonalities in the gentrification experiences of cities situated in different states and societies in his study (2008) of two global financial cities: London and Mumbai. In each case the marketisation of property and privatisation of housing are hallmarks of urban governance, and the local state's commitment to the pursuit of competitive advantage in global capital markets. But as Harris observes the redevelopment of Dharavi, one of South Asia's largest slums, located across the

DOI: 10.4324/9781315312491-8

170　New gentrifiers and emblematic territories

Mithi River from the Bandra-Kurla financial complex, represents a far more exigent experience of displacement than gentrification processes in the British capital.

The scale of industrial decline (then collapse) in London, the growth in the members of the new middle class comprising the gentrifiers and the increasing spatial reach of upgrading across London cumulatively represent distinguishing features of the metropolitan storyline over the last half-century, as elucidated notably by Chris Hamnett, Tim Butler and Loretta Lees—standard-bearers for what is a substantial field of urban scholarship.

Over the 1980s and 1990s a consequential round of gentrification was shaped by the sale of London council flats to occupiers, effectively taking a large volume of housing out of the rental stock, and presaging a shift from predominantly rental to an increasingly owner-occupier housing market in the British capital. Further, even a cursory examination of residential listings discloses cases of 'former council flats' available for purchase in the present: an online site disclosed 7,713 former council flats available for sale in December 2020 (https://www.new.homes.co.uk; accessed 3 December 2020).

London's global status as (together with New York) primary centre of intermediate banking and finance, comprising affluent contingents of business executives, board members and specialised intermediaries, has created a home-grown contingent of elites imbued with the power to compete effectively for housing in prime residential districts.

London's positionality in key debates and discourses

London has served as a key site for debates among scholars and policy-makers concerning the relative merits of occupational change and (alternatively) rent or value gaps in property markets of the city in explaining gentrification processes and outcomes. Arguably the terrains of contestation have shifted to a degree, as circumstances change, and the mix of factors shaping, upgrading in the residential spaces of the city evolves.

The argument stressing labour market change as principal factor was surely compelling over the final three decades of the twentieth century, as the collapse of industrial labour and working class livelihoods provided the initial conditions for gentrification, with members of an ascendant new middle class the prime movers in the upgrading experience, explicated in David Ley's *The New Middle Class and the Remaking of the Central City* (1996), and Chris Hamnett's study of *Unequal City: London in the Global Arena* (2003).

Upper-tier financial and business classes continue to shape gentrification in the current era, and retain a place among the lead actors in the appropriation of residential stock situated within the resonant cultural landscapes in the city. But diverse agencies engaged in the London property market, situated institutionally within the larger financialisation trajectory which promotes higher property values and profit margins, incorporates a key maxim of the rent-gap argument, which gains traction within the gentrification discourse.

The prime movers of twenty-first-century gentrification on the supply side in London are the complex assemblage of firms which service the wealthy clients who constitute key actors on the demand side. London's property sector, comprised of developers, estate agencies, architects, design professionals, marketing companies and a host of other intermediaries, represents a first-order factor of upgrading and dislocation in the British capital.

Gentrification in London: An outline of themes and narratives

In this chapter I address complex relations between gentrification and the reproduction of space, place and territory in London. These experiences reflect the larger storyline of affluent groups and individuals exercising the power of capital within the spaces of the metropolis to their advantage: a defining continuity in the British capital over extended history.

The narrative in this chapter includes both the structural features of upgrading at the metropolitan and zonal scales, as well as localised experiences and contingent actors across London. I situate the idea of 'neighbourhood' as localised unit of study which takes in compact residential areas characterised by proximate social interaction. 'Communities' encompass distinctive ethnic and socio-economic signifiers, and allied cultures of consumption, occupying specific districts in the metropolis, and in many cases are also connected with diasporic networks globally.

Relatedly 'territories' in the gentrification narrative comprise urban fields experiencing penetration from new groups and actors as consequences of larger economic shifts: aspects of 'territorial financialization' (Chapter 3), and including in the London case the concept of megaprojects which are targeted to affluent markets (Chapter 4).

I offer description of selective aspects of social upgrading in contemporary London, with a view to demonstrating effects of both established cohorts and emergent actors within the capital's residential landscapes. The overall tendency takes the form of an intensification of upgrading experiences within established districts of Central London, and in selected communities situated within former suburbs. These include notably Highgate, which exemplifies the spread effect of 'Alpha territories' associated with twenty-first-century gentrification.

I present discussion of variegated forms of social upgrading in London as a means of demonstrating emergent tendencies in new terrains of change, including a narrative of gentrification associated with the displacement of a traditional market in Tottenham, North London. I then address upgrading processes associated with the development of major new football stadia in London—an understudied phenomenon which links capital relayering and the territorial rebranding associated with English Premier League (EPL) football, with new-build housing markets. I then offer an account of new-build gentrification in the Victoria Dock area.

Chapter 8 closes with a discussion of tendencies which represent forces for gentrification at lower registers. These include the complex role of students in

172 New gentrifiers and emblematic territories

upgrading experiences in certain neighbourhoods, where a segment of international students especially may hold sufficient family wealth to insert themselves in localised housing markets; and, second, the role that short-term rental arrangements facilitated by Airbnb and similar online services play in rent inflation, and the removal of stock from the private rental market, observed in global cities such as London and New York.

The evolving field of upgrading in London: New actors and territories

In this century continuity in the gentrification narrative within London includes notably the influence of wealth in shaping the social geography of the city, but also new factors that have added to the force of upgrading and dislocation. There are new gentrifiers, and emergent territories of gentrification in London, as well as extra-territorial effects as low-income groups and individuals working in London are forced to consider residency beyond the metropolis, obliged to negotiate the difficult trade-offs between inflationary rents and higher costs of commuting.

Tendencies include change in London's occupational structure, and more particularly the emergence of extravagantly remunerated professionals and managers who possess far greater resources than their predecessors in upgrading narratives of the last century. These are described in Tim Butler and Loretta Lees' (2006) exposition on the 'super-gentrifiers': elite cohorts producing new rounds of price inflation and upgrading in established neighbourhoods.

New social groups have entered the housing market in London, including a substantial number of transnational elites seeking a residential address in the capital, for reasons which include the social status afforded by the most elegant districts, access to cultural goods and educational opportunities for family members, and, for some, a political haven in a society in which the rule of law in principle obtains.

Change in London and other global cities is also associated with an ancillary upgrading of retail industries, with the decline of traditional high street retail in favour of upscale shopping and dining experiences for affluent consumers and, increasingly, purchases conducted online: with the latter potentiated by the 2020 Covid-19 pandemic which among other deleterious effects acts to suppress personal visits to shops and cafes.

There is also interdependency between what Andy Pratt and I have called 'industrial gentrification' and social upgrading. The rise of digital technology as production modality has led to the emergence of 'new' industrial districts in London, as described in the previous chapter. Affiliated occupational groups in the technology sector have dislodged earlier arts-based creative enterprises, and producing higher-remunerated managers and professionals, thus pushing the boundaries of social upgrading further from Central London.

The complexity of social upgrading processes in London is also manifested in what Antoine Paccoud (2016) describes as 'buy-to-let gentrification'. As Paccoud explains this form of development has been enabled by profit margins associated with new-build housing in the market, and the effects of neoliberal policies which have in the aggregate underpinned the ascendancy of the private rental market this century.

Paccoud's argument effectively meshes two theories traditionally viewed as oppositional: Hamnett and Randolph's (1984) value gap and Smith's rent gap (1979). More specifically the value gap explanation posits that tenure shifts are a reaction to a 'profit maximizing opportunity' (Hamnett and Randolph 1984: 277), and works 'to attract renters to disadvantaged inner city areas which have historically not been appealing to owner occupiers', which in London tends to mean pushing the gentrification frontier further out in East London, and in South London in areas more distant from the river.

Culture, the arts and the gentrification aesthetic

Influential accounts of gentrification, including notably the work of David Ley, emphasised the saliency of culture as critical motivation for gentrifiers, situated especially within the postindustrial terrains of the inner city (Ley 1996). A cultural aesthetic comprised of heritage building types and referential urban histories appealed to the sensibilities of the pioneer gentrifiers. This locational tendency was also shaped by a rejection of the perceived cultural conformity, landscape sterility and social ennui of suburban districts which developed in the postwar city.

In London we can readily identify critical sites of gentrification which conform to a model of upgrading as response to the aesthetics of place and territory, viewed (and experienced) through connection to landscape and the built environment, and to rich amenity *in situ*.

In the referential case of Barnsbury, London Borough of Islington, the initial wave of gentrifiers in the 1970s have been followed (and to a degree supplanted) by the super-gentrifiers: a privileged contingent responding to the place-signifiers of earlier groups who had themselves displaced working class populations of this area of Islington, and including professionals employed in medicine, the law and higher education.

The steep gradient in house prices and rents shaped by the super-gentrifiers in this century, following the initial upgrading experiences of the 1980s and 1990s, is borne out by the experience of Waterloo Terrace, situated to the west of Upper Street and south of Barnsbury Street (Figure 8.1). A flat in Waterloo Terrace which sold for £560,000 in 2000, following a quarter-century of multi-stage gentrification, now has a sale value of an estimated £2.1 million, according to Zoopla, a British property firm (https://zoopla.co.uk/house-prices/London/ Waterloo/Terrace/; accessed September10 2020).

The nature of the consumption landscape of Barnsbury and its environs bears witness to the serial experience of upgrading in this instructive case. The mix of upscale restaurants, cafes, wine bars and coffee shops lining Upper Street as it

FIGURE 8.1 Waterloo Terrace, Barnsbury, London Borough of Islington

winds southward toward The Angel reflect the tastes of an affluent population possessing sufficient discretionary income to allocate to consumption preferences in a corridor shared with visitors and tourists (Figure 8.2).

Reinforcing this affinity between consumers and retailers are stores which reflect the tastes of gentrifiers, including artisanal food shops, bespoke tailors, home décor and paint stores and numerous print shops and galleries: each enabling choice for discerning consumers. Upper Street offers a rich experience of aesthetic consumption and conviviality, in contrast to stores offering basic provisions for working class families, and the proliferation of pubs catering to industrial workers, characteristic of this area of Islington up to the 1970s. Upper Street presents a coherent imaginary of elevated taste and aesthetics of consumption for the latest of successive inflows of gentrifying populations in this iconic territory of the capital.

New industrial districts, localised capital flows and social upgrading

Occupational change accruing from industrial restructuring at the national and metropolitan levels has decisively shaped social class and housing markets over the final three decades of the last century, experienced at the community and neighbourhood scales. In the present century the growth of cultural quarters and

FIGURE 8.2 The aesthetic consumption-scape of gentrification: Upper Street, Islington

new industrial districts comprised of firms engaged in digital production and communications innovation has shaped an important gentrification experience.

London, along with cities such as New York, Chicago, Milan and Berlin, represents an instructive exemplar of this contemporary form of upgrading in the city. While there are aspects of contingency from place-to-place, the upgrading tendency of technology-based enterprise in the city is associated with the price points (and profit potential) of products and services; relatively high wage and salary levels of technology workers, including in many cases generous performance bonuses; and the higher-margin consumption amenities associated with these professionals.

Rachel Granger has undertaken work on the linked themes of capital flows, spatial change and the rent gap in global cities, informed by her work in the London and Chicago cases. Here relational dynamics within what Granger (2015) terms the primary circuits of production, secondary circuits of the built environment and tertiary circuits of science and technology are played out in different ways in specific cities. Granger invokes the literature and analytical manoeuvres around 'gentrination', which take the form of upgrading and dislocation which follow capital flows and 'fixes' mediated by local factors, to explain change at the district level in cities.

Granger observes that in Chicago the financial crisis of 2008 and its aftermath produced massive devaluation of property, followed by speculation, in select areas of the city, although not observed in surrounding areas of the metropolis. Here we can acknowledge Richard Lloyd's study (2006) of Wicker Park in North

176 New gentrifiers and emblematic territories

Chicago, where a localised gentrification experience was observed in the early years of the present century, shaped by a resonant landscape of cultural workers and related consumption amenity. In contrast the South Side of Chicago largely remains a community of entrenched poverty, shaped by unemployment, disinvestment and racism directed against the majority Afro-American population.

Unlike the more spatially limited experience of upgrading in Chicago, and more like the broader and deeper New York storyline, the 'increase in the visitor economy, education and digital media activities have created a new built environment and source for investment flows' (Granger 2015 ppt. presentation: Chicago AAG meeting) in London. In the preceding chapter I offered an outline of the successive phases of upgrading in the Hoxton and Shoreditch Triangle, shaped in part by sequences of higher-value industries and labour situated within this iconic district of East London.

There is of course a larger and more extended trajectory of upgrading within the City Fringe and the revalorised districts of inner north-east London. The 32 hectare Broadgate project initiated in the closing years of the last century, developed by British Land on the northeastern perimeter of the City, effectively extended both the field and gradient of upgrading on the City Fringe.

Pressures for upgrading are realised at the district and community scales in London, expressed in landscape change, higher-margin enterprise formation and the influence of new social actors. A vivid example of the cumulative effect of social, industrial and landscape change is Calvert Avenue, situated between Arnold Circus and the Shoreditch Triangle, separated from Rivington Street by Curtain Road.

In the Victorian era Arnold Circus and its vicinity was one of the most socially deprived areas of East London, rife with poverty, crime and homelessness. Investment in quality housing in Arnold Circus in the nineteenth century in this pioneering public estate succeeded in alleviating the effects of deep poverty for many residents. The housing units and flats situated within the Arnold Circus estate present an attractive and architecturally coherent landscape in this area bordering Hackney and Tower Hamlets (Figure 8.3). The larger area has experienced a steep gentrification experience, although the London Borough of Tower Hamlets still manages about 500 flats in the area.

The upgrading of Calvert Avenue offers evidence of another rescaling of the gentrification narrative. A site visit in the fall of 2018 disclosed evidence of a comprehensive upgrading of Calvert Avenue, replete with craft production, bespoke fashion outlets and artisanal food and coffee shops (Figure 8.4).

In conversation with several of the staff employed in shops along Calvert Street (September 2018), I learned that this micro-space of the revalorised inner city was perceived as a distinctive community of artisans, craft foods specialists and sales staff, adjacent to the Shoreditch Triangle, but for some at least exhibiting a separate identity. Proximity to Arnold Circus was cited by several as an advantage, both in terms of custom, and its value as territorial signifier.

While proximity to Arnold Circus was valued by shop employees for its architectural values and social identity, Calvert Street in fact lies within Shoreditch and

New gentrifiers and emblematic territories **177**

FIGURE 8.3 Landscape of housing estates, Arnold Circus, Tower Hamlets

FIGURE 8.4 Artisanal production, retailing and consumption, Calvert Street, Hackney

the Borough of Hackney, rather than Tower Hamlets. It seems logical to include the upscale businesses situated along Calvert Street within the ambit of upgrading shaped by the serial episodes of restructuring in Shoreditch, and more specifically the increasing presence of business and digital technology clusters which represent the most recent episode in the remaking of this iconic district of East London.

Elite gentrifiers in 'alpha territories': Transnationalism and financialisation

A distinctive marker of the trajectory of upgrading over the present century is the emergence of super-gentrifiers, comprising groups and individuals who possess sufficient resources not simply to compete for desirable properties situated within upscale neighbourhoods of the capital, but also to displace existing residents who number among the merely affluent.

This experience of the elite gentrifiers displacing existing residents, many of whom had themselves succeeded in dislodging middle class householders in the closing years of the last century, is observed in Central London neighbourhoods such as Chelsea and Kensington, and Highgate, an elite suburban village comprised mostly of Georgian buildings, situated at the north-east corner of Hampstead Heath (Figure 8.5), and which has been subject to penetration by wealthy transnationals in recent years.

FIGURE 8.5 Highgate Village, London N6

Wealthy elites elect to invest in alpha territories not solely to enjoy the rich historical ambience and upscale retail and consumption amenity of districts like Belgravia, but also to access cultural capital *in situ*, and to embellish personal and family status. Local authorities have in many cases accommodated the preferences of wealthy residents for additional space beyond that normally permitted. Notably allowances for extra density have been in some cases granted in the case of the infamous 'iceberg houses' in Central London, in which wealthy property owners have been permitted by borough staff to extend multiple levels below grade within the surface floorplate.

This experience recalls earlier examples of elite interests lobbying for extra building allowance in privileged districts of London, exemplified in the case of the 'Babylonian flats' described by Richard Dennis (2008), where builders lobbied the Metropolitan Board of Works in the 1880s to grant extra height and floorspace.

This form of super-gentrification presents a challenge to traditional ideas around upgrading and dislocation in London as identified by scholars over the past four decades. As Luna Glucksberg (2016) has asked: can comfortably affluent homeowners electing to sell their property to wealthier groups and individuals be regarded as 'displaced', while vacating their premises with (often) several million pounds accruing from the sale of their home?

It seems sensible to situate these experiences of the super-wealthy displacing the merely affluent within London's alpha territories as property-based expressions of financialisation, described in Chapter 3, rather than 'gentrification' in its more stringently normative sense. After all, affluent homeowners electing to sell their homes to wealthier individuals possess more agency than the numerous 'middle' class cohorts who struggle to find purchase in the London housing market, and possess far greater resources than working class groups and individuals coping with exigencies of life in an era of government austerity.

Gentrification beyond the inner city: Frontiers of upgrading in London

For much of past half-century the gentrification narrative (and discourse) has been situated within the postindustrial districts and former working-class neighbourhoods of the central and inner city. These accounts reflect the geography of upgrading which incorporates supply side factors in the form of culturally resonant territories and the built environment of the postindustrial city, and the cultural tastes and associated locational preferences of the new middle class.

There are continuing sagas and rounds of gentrification within the central and inner city districts of cities, as new actors enter the storyline. But the present century has also seen important new upgrading tendencies situated within less traditional domains of the metropolis, including hitherto understudied frontiers

180 New gentrifiers and emblematic territories

of gentrification location beyond the inner city. In this section I offer succinct accounts of a selective range of experiences of gentrification in London beyond its familiar habitus.

Frontiers of upgrading in the metropolis: Gentrification sites in South London

The most influential sagas of gentrification in London take place in diverse territories north of the Thames, notably original areas of upgrading within inner north-east London, and including the emergence of 'alpha' territories of super-gentrification in districts like Chelsea, Kensington and Highgate. The south bank of the Thames has also experienced upscale redevelopment, including notably pioneering sites like Butler's Wharf and, more recently Nine Elms and Bankside.

Distinct from these urban megaprojects situated in prime locations, South London has emerged in the twenty-first century as an increasingly important territory—or rather multiple territories—for gentrification. As the price escalation frontier for housing in London north of the Thames shifts ever-further east, a substantial number of those seeking a London residence are motivated to search out possibilities in South London, particularly in areas well-served by public transit. A number of these neighbourhoods present a striking contrast to the narratives of poverty and deprivation of the 1970s and 1980s in places like Brixton, then afflicted by civil disorder shaped by racism, social discrimination and exclusion, lack of employment opportunity and underinvestment in local education and social services.

The redevelopment and social upgrading of South London districts includes communities such as Peckham, Clapham and Brixton, and carries with it cultural signifiers which include amenity in the form of restaurants, cafes, bars and retail stores, as well as new and refurbished living spaces. There are vibrant communities comprised of established immigrant populations dating from the 1960s and 1970s, as well as more recent inflows of professionals and cultural actors who have collectively refashioned (and re-imaged) prominent town centres within South London.

Major thoroughfares, such as Peckham High Street, present especially lively pedestrian traffic flows, replete with both traditional and contemporary retail outlets, as well as sites of worship and assembly for South Asian and African resident populations. These evoke the spirit of Timothy Shortell's idea of 'everyday globalization': areas of identity formation and recognition, localised capital flows and services provision which I describe in the following chapter.

Lavender Hill represents a resonant example of the cultural aesthetics of social upgrading in South London. Far from the louche territorial character projected by the 1960s film the Lavender Hill Mob, the streetscape bears clear evidence of the heightened tastes and cultural signifiers associated with gentrification observed in other areas of the metropolis. These include (Figure 8.6) artisanal

FIGURE 8.6 Consumption landscapes in Lavender Hill, South Battersea, Wandsworth

food products, bespoke tailors and upscale cafes, catering to visitors as well as the local population, endowed with sufficient resources to express cultural preferences through discretionary spending.

A site visit (September 2017) on a Saturday morning disclosed flows of shoppers and other visitors patronising the shops, stores and cafes along the thoroughfare, facilitated by bus service which enhances public accessibility.

Lavender Hill presents as an aestheticised exemplar of tasteful consumption not unlike (at a lower register) Upper Street in Islington, evoking the travelling modality of upgrading as more districts of the metropolis are brought into the ambit of upscale redevelopment. But behind this upbeat narrative of place-remaking it seems important to acknowledge that the remix of social groups in the area in this century presents another instance of residential property effectively taken out of the portfolio of living space available to lower-income Londoners.

'Regeneration as gentrification': Tottenham's 'Latin Market'

A key marker of gentrification in London (and other global cities) concerns the variegation of sites, actors and specific issues in play. In London controversy has been associated with redevelopment pressures on markets, both in the cases of wholesale markets which comprised key elements of the capital's economic geography over periods of extended history, including Covent Garden in Camden.

182 New gentrifiers and emblematic territories

Pressures exerted by developers on the historic Spitalfields Market comprised a key marker of the discourses of postcolonialism and gentrification in Jane M. Jacobs' influential study *Edge of Empire* (1996)

The makeover of Spitalfields into a contemporary site of artisanal goods exhibition and sales, mainstream retail and mid- to upper-level consumption hasn't pleased everybody. But the Market continues to contribute to the liveliness of an area which has undergone successive in-migration and socio-cultural relayering over the past three centuries.

Typically the terrains of contestation over public market preservation in London incorporate a complex of array of social actors and institutions, including agencies of government. In the twenty-first century there can also be a greater involvement of multiple publics, including those directly affected, as well as social media campaigns, crowdfunding to bolster resources available for contesting market and institutional forces, and in some cases appeal to external bodies and councils. Dialogues are established with a growing number of partners and interested third-parties in the fight against the forces of capital and redevelopment.

The proposed demolition of a well-known market in Tottenham, North London, brings competing interests among these diverse interests, actors and institutions to the forefront. Plans to demolish the Pueblito Paisa Market, situated proximate to the Seven Sisters Tube Station, has served to mobilise community resistance since 2008, while bringing into the fray a diverse set of constituencies, and generating a debate which reflects both widespread and more specific concerns.

For their part, the prospective developers, Grainger PLC, have offered what they term a 'generous package of support' for the market vendors, including relocation assistance, so that the site can cleared for the construction of 196 new homes and a shopping plaza, as reported in *The Independent* (21 November 2017: 'London market demolition triggers UN investigation into area gentrification'). Lizzie Dearden, writing for *The Independent*, sketched the contours of a debate which transcended the usual contingents of local interests to engage the Mayor of London (initially Boris Johnson, and more recently Sadiq Khan) and the United Nations, specifically the UN Working Group on Business and Human Rights.

While the Pueblito Paisa Market is located in a less central London location than the Spitalfields Market which provided vital discourse material for Jane M. Jacobs, it raises larger questions than its modest scale (61 vendors in 2018) might imply. The debate highlights concerns about the positionality of local government, with some members of the council making the case for the logic of larger regeneration goals in an area of low investment and underemployment. Market vendors, though, bring to the fore an argument that Pueblito Paisa represents a unique market and meeting place for Latin American immigrants.

These concerns are also given expression in the statement of the UN Working group, issued under the official imprimatur of the Office of the High Commissioner for Human Rights, and involving the engagement of the Special Rapporteurs and Working Group: elements of what is termed the Special Procedures of the Human Rights Council. Special Procedures, the largest body of independent

New gentrifiers and emblematic territories   **183**

experts in the UN Human Rights system, is the general name of the Human Rights Council's independent fact-finding and monitoring mechanism.

Special Procedures mandate-holders are independent human rights experts appointed by the Human Rights Council to address either specific country situations or thematic issues in all parts of the world. Following a report by Surya Deva, chair of the UN working group, the High Commissioner's statement concluded that closure of the market and ensuing demolition 'would result in the expulsion of the current residents and shop owners from the place where they live ... and would have a deleterious impact on the dynamic cultural life of the diverse peoples in the area' (News Release: 'London Market Closure Plan Threatens Dynamic Cultural Centre': Geneva, 29 July 2017).

The extended debate concerning the redevelopment of the Pueblito Paisa market serves to bring together multiple strands of the upgrading narrative in London, including diverse platforms for the voices of local community members, coalitions of activists advocating for the preservation of an important cultural space in the Seven Sisters district, and the dilemma confronting local authorities aspiring to reconcile local cultural values with the interests of capital.

Capital projects and 'new build' gentrification in London

Serial gentrification and new episodes of upgrading are played out within established communities and neighbourhoods in London, while spread effects in emergent spaces and territories of redevelopment are observed. In this century major projects are implicated in emerging narratives of upgrading across the metropolis. Some cases take the familiar form of more affluent groups and actors displacing social cohorts with lower endowments of capital, including the very large numbers of lower-level service workers, while others exhibit forms of indirect upgrading associated with the high price points of local consumption.

Stadium development as centrepieces of local area regeneration and development

The recent redevelopment and upgrading of major sports arenas represent storylines of new-build gentrification in London. The New Wembley Stadium in the London Borough of Brent, and the Emirates Stadium in Holloway (Arsenal FC), have each stimulated the development of high-density residential projects within their precincts, incorporating upscale services and consumption amenities. New football stadium construction for London clubs is enabled by the massive revenue streams associated with the EPL in the form of domestic and international broadcast rights, as well as revenues from ticket holders and membership fees and ancillary consumption amenity on site.

Kathryn Hopkins, writing in *Mansion Global*, has disclosed the inflationary effect of investment in football stadiums in London ('in London, you'll want

184 *New gentrifiers and emblematic territories*

to live near these premier league stadiums': http://www.mansionglobal.com/articles/28103-in-london-you-ll-want-to-live-near-these-premier-league-stadiumsmod=mansion_global_articles_en_wsj_home), establishing the inflationary effects of co-location, and introducing new stadia as inducements to wider property inflation in proximate sites.

Planning for major new stadia typically includes infrastructure investment *in situ*, the provision of public services and transportation upgrades. Capital projects are situated within local regeneration programs and programs for public realm enhancement. For local authorities, large-scale redevelopment projects in the form of condominiums close to major stadia serves to justify public investment in transportation infrastructure, while generating revenues to assist in the capital and operating expenses incurred, and stimulating retail and other forms of consumption.

Sites proximate to the new Wembley stadium in Brent have been subject to redevelopment, favouring high-density housing, and including the provision of community recreation and meeting facilities, private housing incorporating what one project hoarding (observed fall 2017) described as 'stylish apartments', as well as student housing for Middlesex University (Figure 8.7).

The profile of residential development proximate to the new Wembley Stadium, while including public accommodation and community amenity, represents in the clusters of condominiums and apartments a significant extension of new-build gentrification in London, and an upgrading experience within this traditionally working class precinct of the capital.

Capital relayering and social upgrading in the Victoria Dock

In London gentrification at the metropolitan scale also entails a travelling narrative, as developers and allied industries and firms seek out new opportunities in hitherto marginal sites and spaces. In Chapter 4 I described the structure of megaprojects situated within two sectors of the Thames: the established field of redevelopment east of the Tower Bridge, initiated by Canary Wharf, and with projects situated in the Greenwich Peninsula and further downstream; and a relatively new theatre of redevelopment in west-central London, and comprising the Battersea-Nine Elms, Chelsea Barracks and Lots Road-Chelsea Reach projects.

The Royal Victoria Dock, a critical underpinning infrastructure of London's global city role in the nineteenth century, comprises an extensive territory which encompasses multiple redevelopment sites, replete with infrastructure and installations derived from its earlier port, warehousing and distribution functions. The landscape of the Victoria Dock incorporates heritage assets and historical signifiers that resonate both with developers and residents and a concise rehearsal of selective features of the area is appropriate.

The Victoria Dock and adjacent residential communities, formerly part of the County of Essex, shared with the proximate territories of East London a common social history of industrial labour, deep poverty and social deprivation. Population growth shaped by a proliferation of factories in the area led to the establishment of

FIGURE 8.7 Student housing for Middlesex University, Wembley, London Borough of Brent

the County Borough of West Ham in 1886, incorporating Plaistow, West Ham, Canning Town and Silvertown. West Ham was incorporated with East Ham to form the London Borough of Newham in the London Government Act 1963.

The industrial areas adjacent to the Victoria Dock were among the most polluting of all the manufacturing sites of London, Thameside and the Thames estuary. They included industries considered too noxious for East London, with its burgeoning population, and included high-externality industries such as dye-makers and manufacturers of printer's ink. An important episode in the industrial history of the Dock area was the relocation of Samuel Silver's clothing factory from Greenwich, lending his name to the new community of Silvertown. Other prominent industries in the area included food products, typified by Tate & Lyle (sugar) and James Keiller & Sons (marmalade).

Silvertown has been recast in the modern era of community development as Britannia Village—a distinctive feature of the larger Victoria Dock project landscape. The concept of a village-scale project situated within the extensive territory of the Dock was endorsed by the overseeing Royal Docks Trust, and supported by an engaged community planning process. A redevelopment program over the 1990s included the demolition of older tower blocks which had been built by Newham Council, as they were considered to be of poor design and beyond rehabilitation, with the former residents of the flats relocated.

The initial cohort of residents moved into Britannia Village over the last years of the twentieth century, and a number of institutions were established to support community development. These included the West Silvertown Community Foundation. Britannia Village presents as a compact, attractive community ensconced within the larger redevelopment territory of the Victoria Dock (Figures 8.8 and 8.9),

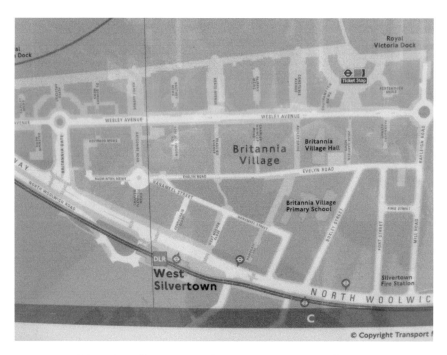

FIGURE 8.8 Britannia Village, West Silvertown, Victoria Dock: Community site planning map

New gentrifiers and emblematic territories **187**

FIGURE 8.9 Britannia Village: Housing form and heritage landscape

complete with carefully preserved heritage features which recall the area's maritime vocation and warehousing role.

On site visits to Britannia Village in the fall of 2017 and 2018, I observed what appeared to be a lively social life within the Village area, including families and small groups as well as individuals, engaged in patronage of consumption amenities including recreation, food and drinks outlets situated in the adjacent quays and dock areas (Figure 8.10).

Britannia Village represents in its compact scale and coherent heritage context an outlier to the overall dimensions and character of the Victoria Dock. The larger Dock area incorporates multiple layers of the larger metropolitan redevelopment experience, including re-urbanisation, regeneration and new-build gentrification, shaped by its scale, landscape features and distance from Central London and the City Fringe which form the essential landscapes of the gentrification narrative in London.

Tim Butler has contributed to the continuing gentrification discourse in his work on the London Docklands, including study of the social implications of development within the upgrading frontier of the Victoria Dock. Butler's scholarship on the new residential landscapes of the Victoria Dock is informed by his work on the more familiar terrains of gentrification in London's inner city, as well as a program of field work in the Victoria Dock.

FIGURE 8.10 Recreational amenity sites proximate to Britannia Village

Butler poses questions concerning the extent to which the new housing in the Victoria Dock represents a model of regeneration or rather a recombinant practice of capital relayering within which multiple elements are present. He rehearses different and to an extent competing interpretations of Docklands within the discourses and rhetorics of new urbanism, capital-led gentrification and suburban development.

A first observation concerns an important gap between new housing and resident social groups in the Victoria Dock, and the sequences of gentrification within inner London. The early fields of gentrification, notably in Islington, and emergent communities of upgrading such as those in South London I described earlier in this chapter, each present for prospective gentrifiers an aesthetically resonant built environment, abundant amenities and a congenial social density. These inner city neighbourhoods, moreover, are situated within comfortable commuting distance of the employment-rich districts of the City and elsewhere in Central London, leaving more time, as Butler notes, for socialising and recreation.

As Tim Butler observes, the new residential tower precincts of Docklands lack these qualities of spatial intimacy, sought after by those valuing the rich social and cultural milieu of the city proper. This has to do with building scale (Figure 8.11), a relative paucity of street-level amenity and of course the much greater commuting distance to Central London. He suggests that for many

FIGURE 8.11 High-density residential units, Victoria Dock, London Borough of Newham

workers in management and the professions, residency in the Victoria Dock presents a form of pied-a-terre occupancy, rather than principal residency.

As individual sites of redevelopment between Canary Wharf and the Victoria Dock are built out, and as more amenity is provided for residents, it's not unlikely that a more architecturally legible and socially textured territorial model can emerge *in situ*. But apart from the lively Britannia Village neighbourhood, the Victoria Dock as a whole lacks the more developed regeneration programming of (for example) the Battersea-Nine Elms project, and the latter's detailed urban design programming across the site as a whole, as well as Battersea's proximity to Central London. So for now at least Tim Butler's assessment of the isolating effects of distance, and alienating consequences of the mega-scale urban design configuration, rings true.

Gentrification below the sight-lines in London

While London's proliferation of megaprojects represent a dramatic manifestation of upgrading in place, I propose to conclude this chapter with selective accounts of trends in housing occupancy which have been in relative terms under-studied: first, a concise description of what some analysts have termed 'studentification', which may impact housing supply and rents in some districts of the capital;

and, second, prospective implications of online accommodation reservation systems for the supply of rental housing in gentrifying areas of global cities such as London and New York.

Higher education and impacts on housing supply and rents: 'studentification' in London

University and college students project a polyvalent role in the social organisation of the city and in the gentrification experience in particular. Students broadly possess fewer financial resources than the average for social groups in the city as a whole, and many struggle to obtain affordable housing, especially in global cities which also function as major international centres of higher education, exemplified by New York, Paris, Milan and of course London.

Students also generate significant benefits for cities, including key roles in the visual and performing arts, patronage of retail and consumption industries, participation in the 24-hour economy of the city and constructive political engagement. Many students contribute to public discourse across a broad range of themes and issues, including myriad forms of online articles, posts and blogs, among other media formats. But students can also contribute to gentrification processes, and more specifically experiences situated in certain communities and neighbourhoods.

In a journal article on impacts of 'studentification' in British cities, Darren Smith and Louise Holt (*Environment and Planning 'A'*: 39: 142–161) identify a complex set of impacts, including 'social' (demographic structure of the local population; resident turnover; supply and demand of services; meaning and symbolism of areas and sites); 'economic' (housing supply and demand for housing, including affordable housing; supply and demand for property, retail and leisure services, spending within the local economy); and 'cultural', including supply and demand for leisure, recreational and retail facilities) (Smith and Holt 2007: Table 8.1, 149). The authors pose the issue of whether students residing in British cities can be considered 'apprentice gentrifiers' based on an assessment of these indicators.

London is a global centre of higher education, and the very large numbers of university, college and private institutional students represent a significant element of the resident population in certain communities of the city. An online issue of *London Higher* (28 November 2017) reported that there were approximately 249,000 undergraduate students enrolled in London's universities, with a further 121,000 graduate students, roughly equivalent when combined to the population of a medium-size town. By discipline approximately 15 per cent of London's university students were enrolled (in 2017) in business administration programs, with a further 12 per cent each in health care and the creative arts and design (ibid).

London's student population includes many enrolled in graduate and professional programs, and large numbers of international students; many of these (but by no means all) have greater access to financial resources than domestic students, and are better positioned to support the high rent levels prevalent in much of London (https://www.londonhigher.ac.uk/ceo-blog/student-numbers-in-london; accessed 23 October 2019).

The complexity of student population impacts at the local level in London are described in a report published in *Southwark Notes*, a local newspaper, where the dual roles of students in both 'regeneration' and 'gentrification' are described, in connection with a proposed project on the Walworth Road. The article includes an observation that 'student housing has emerged as the best-performing asset in the US and UK', with returns in the 12 per cent range per annum.

A case study presented in this article of a conversion of a petrol station for student housing is accompanied by Southwark Council's acknowledgement of local housing demand generated by students attending King's College London and the University of the Arts; but expresses concerns regarding inflationary impacts on housing prices and rents in the area. New student housing on the old petrol station site, Dashwood Studios, was at the time of publication (2019) fully occupied (https://southwarknotes.wordpress.com/local-development-sites/studentification/).

Owing to the very high costs of rental housing in Central London, where many of London's largest universities and colleges are situated, students in the capital increasingly seek accommodation in more outlying districts and communities, notably in neighbourhoods situated within zones 3 and 4 of the Underground. Students seeking residency in these more peripheral communities of the metropolis are to an extent at least in competition with lower-paid workers priced out of the central area housing market, representing a form of spillover displacement.

Online reservations and the rental stock: Exploring impacts of the Airbnb phenomenon in London

London occupies a prominent place within the visitor economy globally, attracting tens of millions of tourists annually. These include very large cohorts of visitors with interests in culture, including live performance as well as the capital's exceptional roster of institutions, public markets and professional sports, as described in Chapters 5 and 6. There are many visitors to London engaged in business opportunity, property investment, or research. Typically these visitors secure short-term accommodation within London's very large hotel stock, or in other cases temporary residency in universities, colleges and other institutions.

Short-term stays in London are available in many parts of London, but are concentrated within the residential precincts of central and inner London. Land and space is expensive in Central London, and these costs are 'priced in' to London's land rents structure. To the extent that a portion of visitors manage to locate accommodation in established residential areas in the form of bed-and-breakfast stays, it has been generally assumed that these numbers are not sufficient to meaningfully impact rents.

192 New gentrifiers and emblematic territories

But the remarkable growth of visitors using online reservation systems to secure accommodation has evidently changed the landscape of housing supply in some parts of London. Homeowners in London with space available for short-term rental can achieve broad access to potential customers via online reservation. Surveys indicate that in the most popular visitor cities such as London traditional hotel and guest accommodation occupancies have held up well, although of course tourism almost everywhere has suffered in 2020 owing to the effects of Covid-19.

But there is some evidence that online reservation systems for tourists have achieved sufficient penetration to disrupt rent structures in areas where previously the volume of visitors was low. There is a tendency for longer stays on the part of some visitors, having wider access to locational choice, and reflecting generational trends which produce larger numbers of retirees interested in longer stays. Research has disclosed an appreciable spatial extension of accommodation available on Airbnb beyond traditional tourist districts, penetrating in some cases communities of low-income residents.

Housing scholars David Wachsmuth and Alexander Weisler acknowledge that short-term rentals expedited by Airbnb and other online reservation systems represent a legitimate concern for community groups concerned about the preservation of low-income housing in global cities such as New York and London. Deploying a research framework aligned with the larger scholarship on the sharing economy, Wachsmuth and Weisler observe that short-term rental streams can contribute to the formation of 'culturally-desirable and internationally recognisable neighbourhoods' (Wachsmuth and Weisler 2018: 1147).

Wachsmuth and Weisler observe that both developers and homeowners can participate in a profitable short-term rental market with relatively low entry and operating costs, creating an emergent form of rent-gap gentrification which can inflate housing values. They estimate that in the New York case between 7,000 and 13,500 housing units have been taken from the rental market since the introduction of Airbnb and other online reservation systems.

There is evidence that Airbnb and other online reservation systems have impacted housing supply and rents in London. With regard first to growth trends, researcher Rowland Manthorpe deployed data from Airbnb's own site to disclose outlines of the penetration of online reservations across the metropolis. Manthorpe accesses common data sources from *Inside Airbnb*, a web-site which tracks trends in London and other cities, which suggest a growth trajectory in London from 215,000 rentals in 2015, rising to about 480,000 in 2016 and to 935,000 in 2017 (Manthorpe 'Wired' 10/24/2018: 'Airbnb is taking over London').

In an article titled 'Airbnb's disruption of the housing structure in London', Zahratu Shabrina, Elsa Arcaute and Michael Batty (2019) examine the geographical patterns of participating establishments in the online reservation market, taking into account diversity of building types and forms of home ownership (but not considering land use types). Their first observation is that Airbnb rentals

New gentrifiers and emblematic territories **193**

in London are positively correlated with purpose-built flats, conversions and flats situated within commercial structures.

The second observation, and posing a worrying normative issue on the face of it, is that Airbnb is associated in spatial terms with private rental neighbourhoods. Shabrina and Batty estimate that Airbnb and other online reservation systems have the potential to take as much as 20 per cent of the rental housing supply from the market, 'further exacerbating the process of gentrification' (Shabrina, Arcaute, and Batty 2019).

Conclusion: Changing fields of upgrading in the global city

London's social geography and residential landscapes have been shaped historically by privilege, notably by income, wealth and access to political influence. These advantages underpinned the formation of the elite neighbourhoods of Central London, typified by the resonant architecture and built form of Chelsea, Kensington, Mayfair and Belgravia, together with affluent suburban villages such as Highgate and Richmond. Over the twentieth century, state commitments to investment in housing for working class populations produced new communities of affordable housing, including clusters of council flats throughout much of inner London and more peripheral boroughs.

Over the last two decades of the twentieth century, these gains in equity in the housing market were undercut by Margaret Thatcher's abandonment of public goods as policy value and the sale of council housing more specifically. An emergent new global London, shaped by primacy in finance, and intermediate banking and the property market and incorporating an influx of wealthy transnationals with the capacity to insert themselves in elite neighbourhoods, brought with it a sharper delineation of privilege within the spaces of the metropolis.

These processes have been fundamental to shaping the lineaments of contemporary London's residential form and structure, which increasingly favours holders of capital. But as I have demonstrated in this chapter, there are other factors in play in the reproduction of place and territory in London, both within inner London and more broadly in outer districts and boroughs. In Central London transnational elites have possessed sufficient capital to insert themselves in the most prestigious neighbourhoods of the capital: a defining feature of twenty-first-century globalisation, observed in cities such as New York, Paris and Hong Kong.

London's inner city continues to perform as prime locus of upgrading in place. Certain inner north-east territories have formed the principal staging grounds both for long-running gentrification storylines (exemplified by the aesthetic sensibilities of gentrifiers in the Barnsbury neighbourhood in Islington), and by a cultural regeneration program in Shoreditch which has been followed by the rise of an economy of digital industries. Further, capital associated with the growth trajectory and market power of the digital economy, and the residency and consumption affinities of East London's burgeoning technology workforce, have generated a new upgrading narrative in London, as Rachel Granger has observed.

These consequential expressions of upgrading within the familiar terrains of London north of the Thames, which have formed the essential materiel for much of the influential scholarship on gentrification, can also be located in certain areas of South London, as my sketch of Lavender Hill in this chapter suggests.

I have demonstrated the increasing complexity of the upgrading tendency in London, involving new actors, emergent forms of capital-led gentrification and the reach of capital in outer London. The rapid growth of online reservation systems for visitors to London, notably Airbnb as well as other networks, represents a worrying trend, as it takes in a growing number of properties concentrated within the private rental districts of the capital.

Some of the most spectacular expressions of upgrading in London can be found in projects located in peripheral territories of London, including high-density development proximate to new stadia such as Wembley and the Emirates in Holloway. The megaprojects described in Chapter 4, situated in the Greenwich Peninsula in the former Docklands, and the Vauxhall-Nine Elms-Battersea megaproject, are linked to Central London by public transit and employment and are promoted by marketing imaginaries which reference local social history.

But residential projects situated in the Victoria Dock suggest not so much a travelling of familiar narratives, but rather a spectacular expression of what Tim Butler describes as 'new build' gentrification in London, where condominium units are often used as pieds-a-terre, rather than as principal residences. We can say in conclusion that London sustains its deep relevance to the gentrification storyline, expressed both in familiar neighbourhoods, and also in the proliferation of new actors, agencies and spaces implicated in experiences of social upgrading and displacement in the British capital.

9

INSCRIPTIONS OF POSTCOLONIALISM AND SOCIAL HISTORY

Class signifiers, persistent localism and vernacular spaces in London

Introduction: Urban change, cultural encounter and social meaning

As we have seen in this volume financialisation in London has taken the spatial form of extensive capital relayering throughout the metropolis. Key processes incorporate sequences of upgrading in established districts and communities, and the capture of new territories for higher-margin development throughout the Greater London Authority area. Capital has touched down in residential communities, with the impress of money associated both with affluent domestic buyers and offshore sources, driving price inflation and new episodes of gentrification.

A parallel 'industrial gentrification' is shaped by technological innovation, globalisation, change in returns to capital and affiliated consumption preferences. Further, the scale of redevelopment has served to occlude (or selectively appropriate) the visual record of social identity in much of London. Capital intensification, technological innovation and new rounds of globalisation have produced new industrial districts and institutional clusters within London.

There are numerous stories of progressive experiments in local regeneration programs, co-operative housing projects, public space provision and green city agendas. There is also advocacy for progressive land use policies which address social needs beyond those of holders of capital. Policy scholars such as Jessica Ferm and Edward Jones (2015) advocate for the retention of industrial land supply in the interests of sustaining a land market with a range of price points, enabling both industrial diversity as well as high-margin specialisation.

London's experience of regeneration incorporates a broader public purpose, including employment opportunities and public amenity contribution. But overall the defining pattern of change in many of London's spaces—financial,

DOI: 10.4324/9781315312491-9

196 Inscriptions of postcolonialism

commercial, industrial, residential, retail, cultural and sporting—has been one of sustained upgrading and dislocation.

Modernity, markets and spatial diversity

A key residual of capital intensification within London takes the form of an overlay of modernity and functional repurposing of space, signalling change in territorial form and social identity. Capital has served to aggrandise elite interests in the reproduction of the built environment. These processes have in turn served to reshape the affiliation of social groups with space and territory: in many cases uncoupling the intimacy of social relations and cultural identity in place.

Projects incorporate gestures to displaced groups in specific sites: embodied in heritage signage and symbols within London, recounted both in official records and informal storylines, and selectively embedded in the marketing of property, as I described in Chapter 4.

The transformative impress of capital across space in London constitutes the master narrative of London's experience of urbanism two decades into the twenty-first century. These processes include much of the old London County Council area, observed in the redevelopment of industrial districts and riverine spaces, and in the upgrading of retail and consumption zones in response to the preferences of affluent markets actors.

The spaces in between: Reading social history and the local vernacular in London

Amid pervasive landscapes of upgrading, new identity formation and social change in London, we can still identify patterns of association which embody continuity, as well as places which have adapted to change at a lower register. These include the 'spaces in between' the built form of capital projects and other sites of capital intensification, where for a time at least there is some leeway for the survival of low-income and marginalised populations. We can also locate in twenty-first-century London places and territories which mesh retail functions, affordable consumption and lively pedestrian flows of mixed social groups.

There are sites and territories in London which exhibit signs of historically layered social identity, and which embody meaning for broader publics as well as for specific users of space. These include notably sites which exhibit legacies of the imperial project and postcolonialism; territories within which signifiers of working class London and traditional industrial labour are still visible; and areas of mixed social identities.

An argument can be made that this diversity can be accommodated and even selectively encouraged in the liberal city of cosmopolitan values, which London represents for many urbanists. But while London doesn't exhibit the marked segregation by ethnicity as does Paris, wealth is an even greater determinant of spatial privilege than is the case in the French capital.

In this chapter I develop the idea of spaces of vernacular culture and social meaning at the localised scale in London which form part of the landscape (and storylines) of contemporary urbanism, informed in part by influential scholars of social change and continuity. In particular I reference Timothy Shortell's concept of 'everyday globalization' (2016), set against the insistent reproduction of space and territory in the metropolis. While Shortell's observations are derived from field work in Brooklyn and Paris, his theory on the cultural embeddedness and social use of space in the globalising city is instructive for the London case

What follows is a discussion of a selective range of sites in London which, I argue, represent resonant spaces of meaning, accommodation and diversity which support the 'city as cosmopolis' ideal, even as the impress of capital infiltrates new districts. My intent is not to establish a comprehensive roster of archetypes, but rather to identify examples of vernacular space in London which embody aspects of social history and contemporary meaning. I conclude with a concise essay distilling what I see as some broader implications of these observations of change at the neighbourhood level in London.

Vernacular spaces within the globalising metropolis: London and its others

Global processes in the metropolis produce a form of hyper-modernity, observed across space, place and territory. The overriding effect of capital and elite social affiliates in the production of privileged space is observed in the corporate landscapes of the central business district of cities, as well as in the price points of high-rise units situated within urban megaprojects.

This view doesn't however leave enough room for diverse social groups which populate (and effectively animate) the cultural life of the global city. Timothy Shortell has contributed to our understanding of the theoretical framing of immigrant groups in the territories of the global city, producing vernacular spaces replete with rich socio-cultural interaction. His starting point is the 'basis of the affective bonds we form with specific places' (Shortell 2016: 1), and how these are shared through formal and informal socialisation experiences.

Shortell acknowledges the divide between sociologists engaged in the study of (on the one hand) larger or 'formal' space-relations, mediated or co-produced through institutions, structures and ideologies, counterposed against the micro-sociological phenomena of 'everyday life' and the 'habitual' (Shortell 2016: 2), although I would argue that there is room for both in contemporary urban studies. As Shortell acknowledges, another crucial divide among practitioners of urban sociology conducting work on the 'everyday' is that demarcated by the 'private sphere and intimate interactions'(Shortell 2016: 3), comprising a principal field of research; and, alternatively, commitments to studying 'how people use public space' (Shortell 2016: 3).

Shortell draws on influential thinkers to explicate the saliency of the everyday (re)production of public space, including Lefebvre's definition of the city as

198 Inscriptions of postcolonialism

incorporating 'plurality, coexistence and simultaneity in the urban of patterns, ways of living urban life' (Lefebvre 1996: in Shortell 2016: 4), and de Certeau's underscoring of the creative aspect of everyday life (de Certeau 1984) in the city and as practiced in urban communities.

Timothy Shortell's contribution to how we read the everyday cultural life of the city emphasises the immigrant experience situated in specific spaces and territories of the city: a particularly vivid expression of the 'urban culturalist paradigm' which acknowledges the 'place-embedded nature of cultural practice' (Shortell 2016: 7). He recognises that while 'collective identities are more likely to be mobile today, they are still embedded' and, further, that '[t]he practice of culture makes a "place" – meaningful urban space – into "home"' (Shortell 2016: 7).

Shortell's story of everyday globalisation in Paris and Brooklyn defines immigrant neighbourhoods as a descriptor for places exhibiting 'signs of identity as mainly belonging to ethnic, religious, and linguistic minorities'; further, these cultural semiotics are not confined to highly localised territories but rather the global city as a whole, referencing Appadurai's concept of the 'ethnospace': a 'landscape of persons who constitute the shifting world in which we live: tourists, immigrants, refugees, exiles, guest workers, and other moving groups and individuals' Appadurai 1996: 33; cited in Shortell 2016: 10).

The original framing idea for Shortell's depiction of everyday globalisation takes the form of what he terms 'spatial semiotics' (Shortell 2016: 12) of immigrant communities. He espouses an essentially structuralist approach, within which '[s]igns of collective identity are embedded in the social spaces of urban vernacular landscapes thorough intentional and habitual performances, both individual and collective behavior, in obvious and subtle ways' (Shortell 2016: 13). This model offers in turn a particularly rich variant of ethnographic study in the globalising metropolis.

London surely represents one of the most instructive cases for the study of how immigrant communities shape local landscapes in global cities. As observed earlier *passim* immigration has effectively remade communities in the capital over the centuries, and more especially the postwar era, with consequential impacts on housing and labour markets, consumption tendencies and the impress of cultural markers. There are important features of differentiation and complexity to acknowledge, reflecting the important variegation of socio-cultural experiences within cities, and more particularly aspects of exceptionalism in the London case.

One aspect of contingency concerns the experience of cultural overlap and relayering shaped by the settlement of immigrant communities in the former working class districts of London, producing new and emergent vernacular landscapes. There are in some cases important residuals of earlier working class communities, expressed in terms of institutions, housing estates, landmarks and other semiotic features.

London's diverse communities present overlays of immigration experienced through history, producing a rich mixture of cultural signifiers within which the contemporary semiotics of neighbourhood formation are juxtaposed with residual traces of earlier groups, as seen in Acton, Kilburn and Tottenham among many other districts of the capital. Then there are the implications of upgrading

Inscriptions of postcolonialism **199**

at the community level in London, as earlier cohorts of immigrants are effectively displaced by more affluent groups, including transnational elites: part of the impactful impress of wealth-holders in the metropolis.

What follows in this chapter are case studies of districts and communities in London which exemplify the formation of local vernacular space in the global city. I first offer a sketch of the lively cosmopolitan cultures animating Peckham High Street, which exhibits a rich palette of international cultures as elucidated in Shortell's model of everyday globalisation, as well as convivial consumption and markers of the area's civic history. Peckham (SE15) forms part of a larger territory of South London which had experienced social unrest in the 1970s and 1980s associated with racial prejudice and discord, and with state cutbacks in social supports as aggravating features.

Next I present a profile of Queensway, a remarkably diverse and lively micro-scale consumption scape located just north of Kensington Gardens, and then a sketch of Arnold Circus and its neighbourhood on the borderlands of Tower Hamlets and Hackney. This latter case demonstrates the juxtaposition of multiple narratives of urban change in the metropolis, including multi-stage gentrification following a century of reformist measures including public housing provision.

My final case takes the form of a discussion of the instructive storylines of industrial innovation, social change and cultural markers in Clerkenwell—a 'city within a city' comprising a particularly rich urban past and insistent upgrading sequences within the City Fringe.

Peckham High Street: Cosmopolitan cultures in twenty-first-century Southwark

In the previous chapter I identified Lavender Hill in Battersea as exemplar of social, cultural and industrial change in London, including clear evidence of social upgrading and gentrification. These features are embodied in the streetscapes, mix of retail outlets, institutions and flows of diverse social groups and individuals which make up this lively South London district.

Peckham High Street, a thoroughfare running east-west through part of the borough of Southwark, together with its proximate sites and spaces, exhibits social diversity and a lively mix of activities within the cosmopolitan city. Although like any urban territory within the globalising metropolis, it must include experiences of competition and conflict exacerbated by the workings of the property market, Peckham High Street suggests an evocation of the ideals of cosmopolis: spaces in the city where there is a measure of acceptance of diversity that extends beyond mere tolerance.

Arrayed along the streetscapes of Peckham High Street are signs of the complex relayering of social groups, cultural mix and business types within this community of South London. Without essaying an attempt at comprehensive streetscape typology, I describe here some observations generated from a site visit in September 2017.

200 Inscriptions of postcolonialism

The streetscape embodies diverse cultural influences, and presents a rolling narrative of the social and cultural development histories of the area. There are significant representations of South Asian and African cultures present, including a proliferation of storefronts of retail activity: clothing and outwear, a rich variety of foodstuffs, cafes and coffee shops.

On my visit on a clear Saturday morning in the early autumn, a lively clientele was patronising these shops and cafes. There were also examples of the ubiquitous services important to immigrant communities, including health and medical services, legal services and also shops which enable wiring money to family and friends within societies of origin.

Observation of the streetscape, storefronts and flows of shoppers and visitors during my site visit disclosed a rich array of services and cultures. Within just a few blocks the possibilities included Morley's Take-out ('specializing in grilled chicken'), at no. 10 Peckham High Street; Muntaha Textiles (no. 13), Barwaqqo women's clothing (no. 13), Iftin Retail (at no. 13, incorporating an Internet café); the Salama Shop (No. 14, household goods); the Cinnamon Bakery (no. 15); the Divine Peace beauty salon (no. 16), the New Dixie Grill (no. 16), Nyxon Properties (no. 17); Kam Foh takeaway (no. 21); the Queen's Kitchen takeaway; 'Computer 4U' (no. 25); 'Bottles' off-licence (no. 27); and Tiwa and Tiwa Fast Food Takeaway (no. 34B) (Figure 9.1).

Bilex Money Transfer (92 Peckham High Street) guaranteed same-day transfers to 'any bank in Nigeria', signalling the importance of the remittance economy in London as in other global cities. Other stores promised ease of money

FIGURE 9.1 Cosmopolitan landscapes of consumption: Peckham High Street

transfers to institutions situated in South and Southeast Asia, underscoring the cultural diversity of Peckham, both in terms of the cultures of places of origin for immigrant populations, as well as the social reference points for those settling in this cosmopolitan quarter of the metropolis.

The cultural diversity of Peckham as exemplified in the fusion of cuisines was described by Marina O'Loughlin in The Guardian online (Friday, November 2016), where she opens by asking first which stores to check out: 'the juice bar that imports its Med-influenced food from the super-hot restaurant next door? The wooden cabin in the middle of a semi-abandoned warehouse doing artisan pasta? The Hong Kong-style brisket noodles or the hole-in-the-wall South American cafés knocking out arepas and cachapas? The record store that does a roaring trade in Tacos?' (https://www.theguardian.com/lifeandstyle/2016/nov/25/snackistan-London-restaurant-review-marina-oloughlin).

As an example of this new wave of inter-cultural vitality, 'Persepolis' at 28–30 offers both a café (nine-and-a-half tables available) and shop, and includes traditional Persian cuisine as well as smoothies and similar favourites and, in a nod to the blending of tastes and eating environments in twenty-first-century London, playfully brands itself as 'Snackistan' (O'Loughlin 2016).

Amid the rich (multi)cultures of Peckham High Street are resonant signifiers of an older London, expressed in current storefronts, products and services and social connections. These include, notably, the Peckham branch of Manze (Figure 9.2), at

FIGURE 9.2 Manze Eel and Pie Shop, Peckham

202 Inscriptions of postcolonialism

the corner of Hill Street and Peckham High Street, reopened following its destruction in the riots of 1985, and featuring eel pie and mash as menu staples: redolent of Victorian London, and still maintaining a clientele seeking traditional fare.

Then there's A. E. Wilson's Cycles, located at 32 Peckham High Street—self-branded as Peckham High Street's oldest retail store, founded about 1870. An account published in the *Peckham Peculiar* ('A free newspaper for SE15') online included a lively storyline of the recent past and speculations about historical connections, including the possibility of Charles Dickens having purchased a bicycle at Wilson's.

Following the death of Norman Wilson in 1995 (91 years of age), the last of the family, the shop was purchased by Steve Hume, 'a proud South Londoner' hailing from nearby Lyndhurst. According to Hume's story, the front window of the shop was smashed in the Brixton riots in 1981, and the shop fell into decline. But with the revival of both the bicycle as mode of transportation as well as Peckham's retail trade Wilson's Cycles has emerged as a thriving business, as reported in *The Peckham Peculiar* (http://peckhampeculiar.tumlr.com/post/167299619110/wheels-of-fortune; accessed 5 March 2019).

No doubt there are tensions and pressures of business associated with the shops arrayed along Peckham High Street. These include, as divulged in casual conversation with shop operators, rising rates and competition from larger nearby retail complexes. And as I have discussed elsewhere the viability of traditional high streets in London is subverted by online marketing and emergent consumer preferences, and of course the pandemic of 2020.

But the very pronounced cultural variety of storefronts, range of goods and services on offer and the busy flows of shoppers and visitors that comprise the quotidian business (and social) experience along this thoroughfare, testify to the vitality of a streetscape reflective of Timothy Shortell's idea of 'everyday globalization' in the twenty-first-century metropolis.

Queensway: Cultural diversity in the borderlands of Westminster and Kensington

London's complex civic history across all fields of human endeavour, replete with layered experiences of industrial, social and cultural change, is conducive to the multi-scalar construction (and reconstruction) of place—only imperfectly captured in the political geographies of local governance, planning and administration. The conditions which facilitate the flows of diverse social groups enable the persistence of places of diversity as expressed in the mix of businesses and population flows which often lie within the interstices of larger territories shaped by processes of capital relayering, globalisation and upgrading.

Areas in London proximate to mainline rail terminals and principal Underground stations have traditionally incorporated terrains of food outlets and shops, with retailers capitalising on the density of passenger flows to generate business and sales, and projecting social vitality at a local scale. Given London's status as the 'railway town' par excellence, with its 13 stations, and the density of Underground and tube stations pervasive within much of the metropolis, these interchange points comprise important sites of daily social interaction.

Within many of the mainline train stations, there has been a spectacular upgrading of consumption offerings, from the basic beverages, meat pies and sandwiches of an earlier era, to more cosmopolitan (and accordingly more expensive) food choices, including French and Italian cuisine, wine bars and the like: a cultural shift from food items as subsistence and convenience, to an experience to be enjoyed in a convivial setting.

There are as well as expensive clothing and other consumer goods stores, with the Eurostar express rail station at St Pancras a notable example of this trajectory. Impromptu (more or less) concerts performed gratis at major stations add to the conviviality of the experience for travellers, exemplified by the St Pancras 'Station Series', including recitations of classical violin repertoire by Charlie Siem among many other artists.

Over time the changing social and cultural diversity of London and its residents, tourists and other travellers is embodied in the mix of consumption activities proximate to transportation interchanges. There are fewer fish shops, and more international offerings, notably South Asian cuisines; and fewer tea shops, and more coffee outlets, including local chains such as Café Nero and Costa, as well as global brands like Starbucks. There is also a marked generational gesture embodied in the mix of retail, food and coffee shops, in recognition of the very high volumes of young people transiting through London's many rail and tube stations.

Queensway (London W2) represents for many an accessible place of social interaction, affordable consumption and experience. The street extends north from Bayswater to Westbourne Grove, with Kensington Gardens to the south. This short street, which can be traversed within five minutes or so from Bayswater to the site of the Whiteley's Department store at the north end of the street, incorporates a lively and heterogeneous mix of retail outlets, cafes and coffee shops, and projecting a lively and casual ambience. Within this precinct are many businesses, as well as an eclectic set of consumer goods available to the public, and more especially travellers and younger visitors.

This precinct within the heart of Central London is well-connected, facilitated by the Queensway Underground station on the Central Line, and the Bayswater station on the Circle and District Line. The volumes of Underground passengers frequenting these stations creates a substantial flow of prospective consumers over the course of the day, as do the pedestrians transiting the area en route to employment sites, the Park, or the many small hotels situated within the area.

For many visitors Queensway and its shops and eating places offers choice, convenience and affordable products and services. On a visit in the late summer of 2018, I counted a dozen different national and regional cuisines on

204 Inscriptions of postcolonialism

FIGURE 9.3 'Everyday globalization': foodscapes of Queensway W2

offer (Figure 9.3), an evocation of Shortell's everyday globalisation, as well as the ubiquitous Starbucks Coffee anchoring the northern end of the street near the Whiteley's department store site. For a Central London location, these food choices offered an affordable opportunity for many seeking a quick meal before transiting to work, or to one of London's many and diverse cultural and recreational sites.

There is a substantial sub-market for a range of goods and services for tourists, and more specifically budget travellers looking for product price points below the norm in the global metropolis. Many of these are packed into the informal congeries of the Queensway Market (23–25 Queensway), including what a visitor posted online on Yelp! as an 'Aladdin's cave' of mobile phone and computer repair shops, hookah outlets, a tiny Brazilian café and a Russian barber offering haircuts for an affordable £7 (in the summer of 2018). Just around the corner on Bayswater are many more shops, cafes and coffee houses catering to budget travellers and others looking for value and affordability.

Even given the pervasive wash of capital throughout the spaces of the metropolis, it's not inconceivable that Queensway can survive as an enclave of cultural mixing, diverse consumption-scapes and affordable price points. The attraction of Queensway is not solely derived from the prospect of affordable product and services, but is also shaped by the convenience afforded by the dense clustering of outlets, and compact scale and configuration of the streetscape.

Arnold Circus: From Victorian slum to the margins of the innovation district

As we've seen in previous chapters much of East and South-Central London have been drawn into the ambit of capital relayering, industrial innovation and gentrification since the 1980s. The London boroughs implicated in these comprehensive processes of change are sufficiently large as to encompass in important cases multiple narratives of capital relayering and displacement, as evidenced in the exemplary cases of Camden, Wandsworth and Southwark.

As corollary, there is a zonal quality to certain processes and experiences of change, with attendant territorial capture occurring over several boroughs—creating expansive landscapes of change. In this regard the 'innovation economy' extends across multiple sites in Hackney, the City Fringe and part of Tower Hamlets, while the 'super gentrifiers' have infiltrated contiguous terrains in Kensington & Chelsea and the City of Westminster.

The range and depth of capital relayering experienced within a single London borough is exemplified by the experience of Tower Hamlets, historically one of the poorest and most socially deprived local government areas in London—first associated with the slums of industrial London over much of the nineteenth and twentieth centuries, and then with the massive industrial disinvestment, restructuring and loss of livelihoods from the 1970s onwards.

Situated on the northern frontier of Tower Hamlets, Arnold Circus represents a trenchant and instructive storyline of redevelopment, encompassing a history of institutional agency in a saga of slum clearance, new housing provision and serial social upgrading. The most recent narrative takes in the site's increasing integration with the cultural quarter and innovation economy spaces of Shoreditch, situated just across the borough boundary with Hackney.

Taking the storyline back to its contemporary origins, this area of London, popularly known as the 'Old Nichol', was characterised in the middle of the nineteenth century by deep poverty even by the standard of the day, described by reformer Charles Booth as 'the most poverty-stricken in London' (https://friendsofarnoldcircus.wordpress.com/history; accessed 5 March 2019). This judgement takes on greater meaning when viewed against the larger backdrop of the social structure of London which, as Chris Hamnett has observed, 'was like a triangular pyramid with large base and very small apex' (presentation to Green College, University of British Columbia 'The Next Urban Planet' series' 2017).

Terraced housing within the area was widely viewed as a site of deprivation within the extensive working class communities of East London. Possibilities for improvement in the socioeconomic conditions and physical environment in the

206 Inscriptions of postcolonialism

area were shaped both by institutional reform and individual agency. The establishment of the London County Council (LCC) in 1888 as local government for the rapidly growing metropolis provided an institutional basis for progressive action. In 1890 the British Parliament approved legislation enabling housing reform, in *The Housing of the Working Classes Act.* This act of Parliament enabled the LCC's compulsory purchase of land for working class housing, as well as the financing and leasing of new housing property.

The movement for progressive change in the area change was encouraged by a local vicar, the Reverend Osborne Jay, who inspired a major slum clearance and new residential building program: the Boundary Estate project. The motivation was to provide an attractive and healthy environment for working class families in the area, with the principal design features including a central circus with elevated bandstand and with seven tree-lined streets radiating outward from the circus (London Borough of Tower Hamlets: A Celebration of Architecture in London's East End (https://www.towerhamlets.gov.uk/documents/leisure-and-culture; accessed 5 February 2019).

This redevelopment program succeeded in housing significant numbers of working-class Londoners. But there was little room for the most deeply deprived residents, as the new housing was priced beyond their means, and many relocated to Dalston or Bethnal Green. Rather than those mired in deepest poverty, the 'deserving poor' engaged in work were the beneficiaries of the new project: an early demonstration of the potency of dislocation associated with new investment in housing in London.

In recognition of the special heritage value of the area, the London Borough of Tower Hamlets introduced a Conservation Area designation (approved March 2007) for the Boundary Estate Area, including a territory demarcated within the margins of Virginia Road on the north, Swanfield Street on the east, Old Nichol Street on the south and Shoreditch High Street on the west. The area includes numerous listed buildings, with the elegant bandstand featuring a Japanese roof-line dating from 1890 as a centrepiece (Figure 9.4). This conservation designation served both to confer heritage status on the area, including 20 blocks of 5-storey flats, workshops and commercial shops, and also to signal a revalorisation process well underway.

Upgrading continues to shape the design values and social mix within this critical area of inner north-east London, at the interstices of Tower Hamlets and Shoreditch. Calvert Avenue extends from the Arnold Circus bandstand westward to Shoreditch High Street, described in Chapter 8, and represents an eastward extension of the consumption-scape of the Shoreditch Triangle. And as Juliana Martins described in my account of consumption features of the innovation economy in Shoreditch (Chapter 7), the Arnold Circus area has been enlisted as a site of business meetings by entrepreneurs, digital economy specialists and property professionals.

While the social history of the area is complex, this saga demonstrates that the industrial record of Arnold Circus has evolved from the worst of mid-Victorian

FIGURE 9.4 Bandstand, Arnold Circus, London Borough of Tower Hamlets

working class slums characterised by deep deprivation in the mid-nineteenth century, to site of high-level consumption: configured by the area's role as business milieu for actors and enterprises within the innovation economy of the active present.

Clerkenwell: Landscapes of social history, meaning and experience

In a city with a 2,000-year history of settlement, rife with recurrent episodes of innovation, change and conflict, Clerkenwell stands out as a particularly vivid territorial exemplar of human experience. The subject of many books and articles, Clerkenwell encompasses an exceptional record of industrial innovation, immigration and attendant experiences of opportunity as well as hardship.

This district also represents a prime site for absorbing cycles of invention, restructuring and social and cultural change in the world city and globalising metropolis. Clerkenwell has functioned as part of the City Fringe's economy, situated adjacent to the northern boundary of the City of London; but also encompasses within its own territory important sites of industrial innovation and restructuring.

Clerkenwell served as habitus for the two of the most influential critics of the harsh social externalities and class divisions of industrial London in the nineteenth century: Karl Marx and Charles Dickens. The deep salience of each continues to the present era: the former developed a comprehensive theory of

208 Inscriptions of postcolonialism

political economy which continues to influence political thought and social action in the twenty-first century, while the latter's accounts of group and individual struggles in the factory worlds, insalubrious housing and debtor's prisons of an industrial city has lost none of its power to bring to life the consequences of the unrestrained market economy at the localised scale.

The Marx Memorial Library, situated on Clerkenwell Green, stands as both repository of critical texts and readings, available to scholars and other visitors; and a demonstration of Marx's continuing relevance within a globalising city where growing inequality and class divisions provide the materiel for political commentary and action.

In this section I offer a succinct profile of Clerkenwell as archetype of change across multiple fields of human experience and endeavour in London. My emphasis is on selective examples of spatial, territorial and institutional change since its remaking as part of London's cultural revival in the late-twentieth century, as well as some aspects of historical resonance for contemporary social groups: including residents, workers and visitors. Within the larger territory, there is sufficient scope for a rich mix of spatial divisions of labour, albeit with upgrading the dominant trajectory, as well as sites of housing and convivial consumption.

Space and territory in Clerkenwell: Aspects of continuity and disjuncture

Clerkenwell encompasses multiple subareas and sites—each of which is characterised by a distinctive industrial record, built form and social history. Much of the northern part of the district adjacent to the Pentonville Road is comprised of residential and retail uses, while the central area encompasses diverse sites of specialised production and key institutions, although housing is making inroads in the current redevelopment experience.

The southern area of Clerkenwell immediately adjacent to the Smithfield Market and the City of London encompasses high-margin consumption activity, shaped by proximity to the City and its workforce engaged within the financial and business sectors. Farringdon functioned as key City Fringe station within the original Metropolitan Line from the 1860s onward. The additional local retail and consumer market stimulus of the Farringdon Underground reconstruction to accommodate the Elizabeth Line within the larger system, with a projected opening summer 2021, promises to encourage new enterprise formation in the retail and food services sectors.

The area of Clerkenwell south of the Pentonville Road encompasses signifying sites and institutions that reflect both localised and larger markers of continuity and change. Among these is the Clerkenwell Parochial (Church of England) School on Amwell Street. The school has a particularly rich history over its 300 years as an educational establishment in Central London, including visits of Charles Dickens to deliver 'penny lectures' in the 1840s, and a transition in the composition of the student body over the twentieth century. Over the first half of the last century,

Clerkenwell Parochial accommodated large contingents of working-class children resident in Islington. Instruction continued during the blitz, with an improvised shelter onsite to protect the pupils, teachers and staff from air raids.

Clerkenwell Parochial School experienced a transition in the composition of the student body over the second half of the twentieth century, with an increasing share drawn from former colonies in Africa and the Caribbean. This pattern of educational affinity faithfully reflects the findings of Tim Butler, Chris Hamnett and Mark Ramsden (2013) concerning the commitment of immigrant families to secure residency in London which maximises the potential of their children to obtain schooling which emphasises academic achievement. The School's values and practices are respectful of the rich socio-cultural quality of the students and their families: expressions of the spirit of cosmopolis as practiced in this historic institution within the heart of London.

Just south of the School, in Percy Circus, a blue plaque commemorates a seminal political figure of the twentieth century, and another expression of Clerkenwell's history as key site of radical thought and action: 'Vladimir Ilyich Ulyanov Lenin 1870-1924, Founder of the USSR. Stayed in 1905 in Percy Circus which stood on this site'.

Continuing southwards in Clerkenwell, one encounters key spaces which accommodate the economy of consumption and experience, including the lively ambience of the Exmouth Market (Figure 9.5). The Market experienced

FIGURE 9.5 Exmouth Market: Clerkenwell

210 Inscriptions of postcolonialism

regeneration over the 1990s, part of the overall cultural development trajectory within Clerkenwell and other districts in London within the City Fringe. The streetscape is anchored by the Italianate Church of Our Most Holy Redeemer situated in the middle of the block, facing the Exmouth Arms: with the latter an exemplar of an English cultural idiom situated within a diverse consumption landscape.

In the first decade of the present century further refinements to the operation of the Exmouth Market were introduced. These included sidewalk seating and tables for patrons of the approximately two dozen cafes and coffee shops in the market, producing an elevated social ambience. Then in 2006 the Exmouth Business Association approved temporary market stalls in the street, and including a range of food types drawn from (notably) African, Caribbean and South Asian cultures, with prices pitched at affordable levels, and contributing to the multicultural atmosphere of this precinct.

A distinctive feature of the landscapes of consumption in the area is provided by Gazzano's Italian deli and café, located on the western side of Farringdon Road just to the south of Exmouth Market (Figure 9.6). The Gazzanos were members of an inflow of Ticenese immigrants to the Clerkenwell area in the early years of the last century. The residency and tenure of the Gazzano family represents part of both the social history and business development storyline of the area: the deli and café continue to attract a lively clientele of regular business customers and occasional tourists. But what make their experience particularly noteworthy is the development of upper floors of their building for the residential market, enabled by family ownership of the site.

And on the eastern edge of Clerkenwell, the Jerusalem pub on Britton Street offers a congenial atmosphere for designers and other creatives, local residents and occasional business types, with traditional ales as inducement to socialise close to the businesses and residences along St Johns Street. Discussions with the publican (early fall 2016) disclosed a mixed clientele, with locally based artists, design professionals, property market specialists and estate agents and other business groups represented.

Upgrading tendencies in the heart of the global city

The central area of Clerkenwell, situated roughly between Bowling Green Lane on the north and Clerkenwell Road on the south, has performed as site of creative work and cultural production associated with the larger 'arts and culture' turn in London since the 1970s. This sub-area of the larger district of Clerkenwell is replete with narrow, sinuous streets and historic buildings which frame a compact territory of cultural production—for individual artists, artisanal associations and professionals such as architects, interior decorators and fashion designers.

Attractions in this quarter of Clerkenwell include amenities in the form of cafes, coffee shops and wine bars, as well as diverse creative supply firms: forming critical consumption elements of a robust production network in the heart of the district.

Inscriptions of postcolonialism **211**

FIGURE 9.6 Gazzano's Italian Deli and Cafe, Farringdon Road

These resonant qualities of place, including heritage infrastructure, spatial intimacy, professional supply shops and convivial consumption, are also conducive to high-margin enterprises which shape an insistent form of upgrading and dislocation. In this regard my program of field work in Clerkenwell since the mid-1990s

212 Inscriptions of postcolonialism

discloses a paradigmatic trajectory of change, from the initial wave of artists and designers, to professional creative firms and then to corporate clients whose space needs and consumption preferences drive an insistent upgrading tendency.

These changes are registered in the configuration and quality of workspace within the iconic Clerkenwell Workshops, located in No. 31 Clerkenwell Close, in the heart of the district (Figure 9.7). The original conversion of the Workshops

FIGURE 9.7 The Clerkenwell Workshops

to art spaces in 1975 accommodated a lively contingent of artists, artisans and craft workers within 150 workshops and studios: an environment well suited to the needs of a large and growing class of artists and other creatives in the metropolis.

This vocation was truncated in the first years of the present century, with one rupture associated with a general upgrading of inner London's arts scene to a more business-oriented professional design culture, and (relatedly) another with a building conversion directly appealing to design professionals and companies by Workspace Group Plc: owners of multiple properties within the larger City Fringe area. The aim was to promote a more 'contemporary' client base for the Workshops within an area-based regeneration program, bringing with it potential for higher rents.

Building upgrade and rebranding has produced a new mix of tenants for the Clerkenwell Workshops, including professional design firms, consultants and media companies: prefiguring new social divisions of labour on the site. The messaging for prospective tenants includes the lure of 'a homely, yet stylish environment with lots of nooks and crannies …. The building itself has heaps to offer with a delicious on-site café and bar … [and] two fabulous meeting rooms, and a courtyard to enjoy in the summer' (Club Workspace—Clerkenwell Workshops; https://www.spacing.com/location/3393_7; accessed 9 March 2019).

While there is a distinct upgrading of the social milieu associated with the increasingly higher-margin business orientation of Clerkenwell, the southern areas of the district project a particularly lively scene. Repeat visits to the area 2014–2019 disclosed a relayering of businesses along Cowcross Street, proximate to Farringdon Station and the new Crossrail Farringdon station redevelopment, just north of the Smithfield Market.

Here we find Turkish barbers, a restaurant serving authentic Pugliese cuisine, with menu items sourced from the *Mezzogiorno* and upscale drinks places on offer to consumers living or working in Clerkenwell, or transiting the area (Figure 9.8). Once the Elizabeth Line service is initiated, new flows of residents, workers and travellers will enlarge the market base for this southern crescent of Clerkenwell bordering the City of London, stimulating in turn new enterprise and further social divisions of labour within its diverse spaces.

Conclusion: Cultural signifiers and social resiliency in the metropolis

In this penultimate chapter of *Millennial Metropolis*, I have presented case studies of selected London districts and communities which, amid larger and more forceful processes of capital relayering, present signifying aspects of vernacular space and persistent localism in the global city. These cases do not of course represent the full range of local experiences in London but, I argue, are illustrative of socio-cultural accommodation at a lower register.

The larger storyline in London over the past half-century is one of comprehensive change, associated with financialisation enabled by compliant governments

214 Inscriptions of postcolonialism

FIGURE 9.8 Culture, upgrading in place and the aesthetic of consumption: Cowcross Street

and state institutions characteristic of neoliberalism, and the workings of the urban growth machine anticipated by Harvey Molotch (1976).

But as I have demonstrated there is still local agency in London as demonstrated in the quotidian behaviours, interactions and transactions of groups and individuals at the community and neighbourhood levels. As Anthony D. King observes in his *Writing the Global City* (2016), space, form and architectural values are shaped by a complex interweaving of globalisation, postcolonialism and 'contested' notions of modernity: a framing construct which applies in particular ways to London.

Field observations from the sites discussed in this chapter imply the need to be attentive to social change in the modern metropolis at more localised scales. Roy Porter (1994), notably, offered acute perspectives on the changing social geography of London over its tumultuous episodes of growth and change from the industrial city in the nineteenth century, to the socio-cultural implications of the Thatcher neoliberal revolution of the 1980s.

Chris Hamnett, Tim Butler, Loretta Lees and Andrew Harris have each contributed theory and empirical evidence of social class reformation in London in the wake of industrial restructuring, and more particularly the locational preferences and housing tenure correlates of a new services society which includes both a 'new middle class' but also a much more affluent contingent of high earners.

These constitute the broad contours of social change and community reformation in London. I have argued in this chapter though that a textured picture of London's social geography requires recognition of the more localised features of cultural life at the district, community and neighbourhood scales. Study of communities of vernacular culture enlarge our sense of the intimate social connections between the postcolonial metropolis and societies of origin in the global city, and a more nuanced understanding of the use of social space.

The influence of culture in the social reproduction of communities is evidenced in what Timothy Shortell has described as the semiotics of immigrant settlement in the metropolis. In London among other cities these are manifested within institutions, including places of worship and assembly, as well as shopping areas and sites of convivial consumption, particularly evidenced within the food cultures of immigrants.

There are remnants of working class communities, industrial neighbourhoods and allied consumption signifiers in London which have persisted from the late-eighteenth to mid-twentieth centuries. These have been enlisted as cultural markers in redevelopment projects throughout much of the metropolis. The complex echelons of service, food and retail labour observed in each of the districts described in this chapter comprise a 'new working class', replete with divisions of labour shaped by education and skills.

This twenty-first-century workforce experiences challenge in organising for better pay and working conditions, albeit with advantages over previous eras including possibilities of digital communications and political mobilisation, but with the potency of the neoliberal agenda working against the exercise of workers' rights. There are other avenues for forming workers' coalitions in London, including those shaped by ethnicity and familial and social ties, and also the generational propensity of millennial workers for achieving organisational scale through digital media network formation.

216 Inscriptions of postcolonialism

A more critical perspective positions services workers as members of a 'new proletariat' or alternatively precariat, imbued with little bargaining power in markets where the interests of capital prevail. The lively scenes of pedestrians, shoppers and flaneurs experiencing the localised landscapes of amenity in each of our four sample sites in this chapter cannot mask the harsh difficulties of securing sustainable employment and residential tenure within the revalorised spaces of the global city.

The redevelopment experiences of the four districts selected for this chapter present important contrasts in terms of scale, complexity and patterns of social groups, and in their behaviours and experiences. Queensway represents a micro-space of multicultural retail, consumption and experience in the heart of Central London, situated within the interstices of Kensington and the City of Westminster.

The Arnold Circus-Boundary Estate site and Peckham High Street each represent areas of historical layering of social class, ethnicity and enterprise formation. Each now accommodates a lively mix of social groups and cultures, although the proximity of Arnold Circus to the high-intensity innovation enterprises of the Silicon Roundabout area just across Shoreditch High Street imparts a more insistent upgrading narrative in this case.

Clerkenwell is large enough to encompass multiple industrial storylines and social narratives, and in this respect bears comparison to the South of Market Area (SOMA) in San Francisco, where, as Kevin Starr (1996) has observed, each of the markers of urbanism—industrial, social and cultural—are embodied within a single city district, mediated through the filters of restructuring and insistent upgrading since the 1980s.

Clerkenwell and SOMA each experienced inflows of migrants in the nineteenth century; sequences of high-value industrial production regimes and associated skilled artisanal labour in the twentieth century; and increasing social and industrial upgrading over the last decades of the twentieth century, with further episodes of industrial innovation in the present century. SOMA performed for over a century as complex site of industrial specialisation and affiliated labour, as well as comprising a community which absorbed inflows of immigrants from diverse source societies: each of which also comprised elements of Clerkenwell's development pathway over the same historical period.

At the larger urban-regional scale SOMA comprised an important industrial district within the Bay Area's economy of manufacturing, marine industries, machine shops and warehousing and distribution. As in the Clerkenwell case, the decline and eventual collapse of specialised production industries and labour in SOMA produced a legacy of spaces, territories and buildings which have proven conducive to adaptive reuse in the form of cultural industries infused with deeper digital content, and with technological labour. As in the Clerkenwell case, the property market has actively sought to enlist social history and markers of an industrial past, in the reproduction of space, place and territory in this culturally resonant district of San Francisco.

There is also the larger backdrop of civic as well as state-level neoliberalism as forces shaping a harsher social reality in each case, as the liberal cosmopolitanism of both London and San Francisco have been subject to the stringent outcomes of neoliberal political agendas since the 1980s, as Rebecca Solnit has eloquently described (Solnit and Schwartzenberg 2000) in the case of the Bay Area.

Clerkenwell represents a larger territory than the Peckham High Street, Queensway and Arnold Circus areas discussed in this chapter. But what is common to each is the rich mix of historicity in the formation of the built environment, the differential structures of power and agency exercised by diverse social groups and the complex intersections of class and ethnicity played out in space and territory.

10

BREXIT LONDON

The liberal city in a neoliberal state

London as site of change: Agenda and issue framing

The complex interaction of local and global factors which reshape space, place and territory in London forms the central narrative of this volume. There is a resonant empirical quality to spatial change in London, so I have endeavoured to offer a lively portrayal of the quite remarkable extent and character of the production (and more frequently reproduction) of space and its consequences in the British capital. Study of the London case also offers possibilities for contributing to critical concepts of governance in globalising cities and urban systems, and to a deeper understanding of the relational geographies of urban development.

In this concluding chapter I offer a précis of what I believe to be the most compelling themes and problematics addressed in this volume, together with some thoughts on how a study of London can contribute to an enhanced understanding of twenty-first-century urbanism. Following a concise rehearsal of London's value in understanding the larger (and more complex) arenas of comprehensive change in the contemporary metropolis, I relate these to the reconstruction of space, place and territory.

I offer a perspective on the changing role of governance, the state and local agency in the reproduction, management and marketing of space. Here I acknowledge that 'space' is not simply the receptacle for change, nor solely a taxonomical device; but rather in the present global era the staging ground for human endeavour, aspiration and identity formation, as well as critical resource for market penetration and profit accumulation. As metropolitan cities such as New York, Hamburg and Shanghai demonstrate, 'space' in the city is activated as a potent instrument of globalisation, place-marketing and capital relayering.

DOI: 10.4324/9781315312491-10

Next I offer a periodisation of spatial change in London in the late modern era, starting with the catalytic effects of the political mobilisation of space in London circa 1980–2000; followed by two decades of capital deepening and extensification within the larger metropolis. Throughout the discussion I emphasise the utility of space for mobilising development in the city—for capital, social groups and the state. Space is deployed as territorial expression of cultural identity and field of social interaction, as resource for upgrading and value delivery and as key marker of civic culture as deployed in external marketing ventures.

I conclude with a description of spatial taxonomies organised around scale and specialisation derived from the London case. These in turn inform the construction of three scalar models of spatial development, associated with economic development, power projection and social identity in the city, which I characterise as: platform effects; territorial effects; and sites and place-based effects.

London's development trajectory: Governance, agency and state actors

State actors and institutions have been instrumental in the remaking of London over the last half-century. The central state has concerned itself with London as the principal motive site for Britain's redevelopment, modernisation and globalisation, as evidenced in the quite remarkable sequence of capital projects described in this volume. These have contributed both to London's primacy and to the reformation of the capital's space-economy, but have also exacerbated regional disparities in Britain on a much larger scale than that described in Doreen Massey's classic monograph on the spatial division of labour 35 years ago, and acknowledged in her more recent *World City* (2007) monograph.

British Prime Ministers have taken a personal interest in London, starting in the late modern era with Margaret Thatcher's impactful conversion of the Isle of Dogs to the Canary Wharf megaproject, followed by Tony Blair's sponsorship of London's bid for the 2012 Olympic Games. More recently David Cameron's advocacy for a technology corridor linking the post-Olympic Stratford site with 'Silicon Roundabout' in Shoreditch, an episode described in Chapter 7, represents a singular moment testifying to the personal interest of a national leader in the space-economy of the British capital, along with its associated imaginaries of place and territory. Cameron's messaging presents a faint echo through history of Harold Wilson's personal endorsement of the 'white heat of technology' as panacea for the moribund British economy, enunciated at the national Labour Party conference in Scarborough in 1963.

At the metropolitan level a largely progressive and successful governance for London was exercised by the London County Council from 1888 to 1963, followed by the unhappy experience of the Greater London Council: ill-equipped to take on the challenge of governance in an era of massive industrial disinvestment, and caught all too often between the powers and interests of the central

220 Brexit london

government in the form of the Department of the Environment, and the 32 London boroughs as the basic level of local government in London.

The Greater London Council (GLC) was peremptorily abolished by Margaret Thatcher's Conservative Government in 1986. Its successor, the Greater London Authority, led by an elected Mayor for London imbued with greater powers than those exercised by the Leader of the GLC, has concerned itself largely with actively promoting London's global city aspirations, augmented by regeneration programs at the borough level, thus aligning agencies of the multi-level state in this overarching project.

London in the international arena: Europe, world cities and globalisation

The European context for London's development over history is crucial, shaped by flows of immigrants, capital and commodity trade between continental states and societies and London. More recently, the end-game of Britain's imperial project in the postwar era, and the onset of a complex post-colonial trajectory shaped by immigration and settlement in British cities, coupled with experiences of crippling disinvestment and industrial decline, prompted Britain's entry to the European Economic Community (EEC) in 1973—an implicit political acknowledgement that there was much to gain from formal association with important continental polities, economies and societies.

London was a principal beneficiary of this continental association, and was already acknowledged as one of seven 'world cities' in Peter Hall's taxonomy (1966), along with Paris, the cities of the Ruhr and the Randstat, Moscow, New York and Tokyo. No fewer than five of Hall's seven world cities are situated in Europe. If Hall was writing in a similar vein today, he would likely include in his leading seven world cities influential city-regions such as San Francisco and the Bay Area, Hong Kong and perhaps Mumbai or Shanghai, at the expense of some of his European examples.

The collapse of traditional industry in London and other British metropolitan cities and resource regions over the 1970s was an entrée to Margaret Thatcher's neoliberal revolution, privileging finance, property development, other intermediate services and professional labour and managers: the defining cohorts of an ascendant new middle class. The divergent tendencies of industrial decline, and an ascendant services economy produced higher factor and labour productivity throughout much of Britain, but was accompanied by deep social costs in the form of massive employment-shedding in traditional sectors and industries.

London's status as financial capital of Europe was key to its ranking as one of three 'global cities' in Saskia Sassen's well-known text (2001[1991]), potentiated of course by corporate control exercised in key multinational industries, and the global reach of London's intermediate financial sector. Subsequent rankings of global city status, notably those conducted under the auspices of the Global and World Cities network at Loughborough University, affirmed London's global city roles across a markedly diverse field including culture, higher education and knowledge industries and non-governmental organisations—the very definition of what the Global and World Cities (GaWC) terms a 'well-rounded' global city.

While its primacy in finance constituted a cornerstone of its status within Europe and international money markets, London also benefitted greatly from increasingly rich connections and network formation with other European cities and societies, notably in the business, trade, medical, educational and cultural sectors. The integrity of these vital and enriching associations are at risk owing to the delusional Brexit vote of 2016 and its enactment in January of 2020 by a revanchist Conservative government, while Britain's status within a world order increasingly favouring China and other East Asian growth economies, as prophesied in Giovanni Arrighi's *Adam Smith in Beijing* (2007), is very much in question and quite likely parlous. Certainly the 'Singapore on the Thames' trope advanced by prominent Brexiteers as a way forward post-EU is sheer fantasy, as few of the key development conditions underpinning the tropical city-state are evident in the UK.

Divorced from its continental partners Britain will effectively be 'going it alone', in an increasingly competitive world trading environment replete with protectionist states: a far cry from the global mercantilist trade flows that Britain did much to foster and in large part control or manage in the eighteenth and nineteenth centuries, as articulated in John Darwin's penetrating account of *Unfinished Empire* (2012).

Instead the national leadership will have to contend with authoritarian leaders abroad who evince no particular regard for Britain in the negotiation of what are likely to be difficult (and asymmetrical) trade arrangements. London, with its strength in finance, intermediate services, culture, higher education and innovation, supported by its positive global imagery, is better placed to thrive than other British cities and regions, but will share in the national problematics of scale limitations, geopolitical isolation and declining comparative advantage.

Tracking London's development trajectory over time: The remaking of space in a global city

Two decades into the twenty-first century, and four decades into the contemporary era of globalisation forcefully initiated by Margaret Thatcher's neoliberal makeover of the British state, its politics and economy, London presents as a largely successful exemplar of an emergent form of metropolitan city. The British capital combines—not without contradictions and conflict—complex aspects of national economic primacy, global power projection and cosmopolitan urbanism. And in the present century London has garnered the largest shares of new employment growth among British regions in the important financial, property,

222 Brexit london

cultural, technology and higher education sectors, sustaining its national primacy both in established and emerging fields of enterprise.

Periodising London's redevelopment in space: Fin-de-siècle to the twenty-first century

Here I characterise London's redevelopment trajectory over the last four decades: the era of the neoliberal state and the age of capital hyper-mobility. First: the late-twentieth century is viewed as a seminal period of change over London's extended governance and redevelopment history, which I call the *political mobilization of space* in the late modern period.

The period from 1980 to the turn of the twentieth century encompassed the seminal Canary Wharf global financial megaproject; the restoration of metropolitan government in the form of the Greater London Authority including an elected mayor; and completion of the Jubilee Line, linking southeast London and Stratford with Central London. The latter project facilitated the larger redevelopment of southeast London (notably in Southwark) and East London as well as Canary Wharf, enhancing the connectivity of these districts with the established financial, business and state agencies of Central London.

The last two decades of the twentieth century also encompassed a signifying cultural turn for London in the adaptive reuse of the Bankside Power station for exhibition space in the form of the Tate Modern, and a culture-led regeneration of former industrial districts which initiated sequential processes of succession and upgrading within inner city boroughs such as Islington, Hackney and Southwark.

The first two decades of the twenty-first century I characterise as an era of *capital deepening and extensification* within the established zones and districts of London, and increasingly within new development sites and territories. Investment is sourced from a more varied range of international institutions and social actors, and with serial upgrading within the former industrial districts of London a defining feature.

Even a partial list of strategic redevelopment projects in London over the first two decades of the present century would include the massive infrastructural project of Crossrail/HS1 (the Elizabeth Line when open in 2021); the 2012 London Olympic Games; the extension of Thameside megaprojects both downriver from Canary Wharf and upstream from Westminster Bridge; new billion-pound stadia for Spurs and the Arsenal, as well as the adaptive reuse of the Olympic stadium in Stratford for West Ham United's home grounds; the upgrading and relabelling of cultural areas as 'innovation zones', shaped in part by aspirational regeneration programs involving consortia of local government agencies; and major public (and private donor) investment in many of London's museums and galleries.

Internal experiences of growth and change, generally favouring higher-margin industries and professional employment, supported by massive infrastructure investment, have been instrumental to London's development over the past four decades. The comprehensive transformation of space in London, including the deployment of culture and social history in the rebranding of place and territory, has been critical to the formation of cosmopolitan imaginaries which appeal to aspirational cohorts of workers across key sectors, and adding to London's projection within global circuits of trade and exchange.

Amid this pervasive mobilisation of space for capital relayering, we can recognise the persistence of sites and territories which still afford possibilities of vernacular cultures and mixed-use including affordable consumption and experience, as I described in the preceding chapter. London exemplifies in some communities the 'everyday globalization' experience described by Timothy Shortell in his saga of immigrant community life in Paris and Brooklyn.

But there are at the same time communities of deeply entrenched poverty in contemporary London, including disadvantaged households within areas such as Bow in East London, and overcrowded housing estates in places like Brent situated within the outer north-west. In these and other low-income areas characterised by social programs under stress, many Londoners are living a subsistence life in alienating tower blocks, and with few opportunities for younger populations especially.

The growth experiences of the last 40 years have produced a new zonal structure of strategic industries and institutions (Figure 3.1) in London. What this diagram shows is the material expansion of the spatial ambit of development in London, drawing in previously low-density, low-impact sites beyond the central area, while also producing capital relayering, upgrading and dislocation in established areas of the metropolis: an expression of 'territorial capitalism' at the metropolitan scale.

The remarkable expansion of London's space economy and metropolitan territory over the Victorian era, stimulated by new rail systems and the inception of the Underground, stand as the most extensive in the modern era, while the new industrial districts and residential communities of the interwar and postwar periods of the last century certainly represent a consequential experience. But we can confidently recognise the last 40 years as a particularly salient era of redevelopment in the extended history of this 'Millennial Metropolis'.

The mobilisation of space in the global city

Globalising cities have experienced new forms of territorial development in this century, including both established metropolitan cities such as Chicago, Toronto and Barcelona; mega-cities within the growth economies of East Asia like Tokyo, Seoul, Taipei and perhaps most of all Shanghai; and emergent cities of the Global South, typified by Mumbai, Jakarta, Johannesburg and Sao Paulo. There are trend-lines that cut across each, notably sustained inflows of capital, receptivity to multinational development partners and interests and the persistence of informalism within urban-regional economies and labour markets: a feature of

224 Brexit london

Los Angeles and Rome as it is of Accra and Kuala Lumpur, accompanied by the rise of precarious work and splintering labour markets globally.

These development conditions shape growth and change across cities and urban systems, although of course there are quite significant differences, mediated by the quality of governance, public policy values and social capital. London was particularly well-positioned to benefit from the principal growth trajectories of the past four decades: financialisation, enabled and promoted by deregulation and the 'Big Bang' from the1980s onward; the growth of culture and creative industries and labour, from the 1980s and 1990s and into this century; and the rise of the innovation economy, observed in London in the emergence of the digital cluster first in inner north-east London, but also potentiated by the role of leading universities and other institutes, and in the attraction of global players like Google and Facebook.

Allied to these trajectories of growth and change are other important aspects which have shaped London's development experience, including critical supply factors, notably skilled labour, land and infrastructure. But governance, the state and policy values at different scales have also been quite instrumental to London's redevelopment storylines: from international agencies, particularly those within the European Union, to central government and to the complex and multilayered systems of local government within London: each has been implicated in reproducing London over the past half-century.

London's twenty-first-century development storyline, like that of other European cities such as Paris, Milan, Barcelona, Hamburg and Berlin, is shaped in spatial terms by the redevelopment of culturally resonant places and territories within strategic zones and sites of the metropolis. But what's different for London is the scalar quality of redevelopment. Extensive tracts of space have been subject to capital relayering, including not just the compact territories of postindustrial East London, but also the miles of riverine spaces, from Battersea to the Estuary, as well as the many residential communities subject to upgrading both in former working-class communities and, more recently, middle-class districts.

New spaces and territories deployed for redevelopment include south-west London, exemplified by the Vauxhall-Nine Elms-Battersea project on the River, and the messier Elephant and Castle project. A significant regeneration program is underway in Kilburn, situated in the north-west London industrial zone dating from the interwar era—described by Peter Hall as a new factory zone for London a century after the growth of the light industrial zone within the City fringe and north-east London.

Space as register of urban change and field of study: Learning from London

For this section I offer a structure of the meaning of space as a critical dimension of urban growth and change in the contemporary city and metropolis, derived principally from an extended program on research on London, and augmented by reference to other cities, and structured by a systematic disaggregation of functional spaces.

Changes in governance and the state, the globalisation of capital and population flows and innovation in technology have influenced patterns of growth and change in cities: breaking the mould of the mid-twentieth-century city dominated by a compact central business district, industrial districts and residential communities shaped by socioeconomic segregation. First, the spatial patterns of high-margin development exhibited in Figure 3.1 demonstrate the saliency of clustering as a model of growth and change, replete with intricate linkages between firms and specialised labour over a broader territorial expanse beyond the central area.

Mid-twentieth models of industrial location, as explicated in Allen Scott's influential theory (1982, 1988), take the form of industrial clusters shaped by the dense interweaving of input-output relations. A contemporary feature of industrial location is observed in the serial upgrading of cluster content associated with higher-value enterprise, shaped by product innovation and more particularly the market pressures of rents. But the concept of the *industrial district*, reconfigured by digital technology and proximate consumption amenity, retains its saliency in the present, as the London case demonstrates.

A second important principle of spatial development as demonstrated by our London case concerns the *axial growth effect of major transportation corridors*. As Mike Raco has observed, while the collapse of the Port of London and riverside industries for a time represented disinvestment and an evacuation of industrial firms and labour along the riverine corridor, the Thames still embodies special resonance in the British capital's pattern of growth and change, observed in the pioneering Canary Wharf project in the last century, and a spectacular sequence of megaprojects both downstream from Tower Bridge and upstream from Westminster—each imbued with impactful cultural signifiers. Major motorways such as the A40 still have the power to shape corridors of investment and economic activity.

A third critical feature of space in the London case concerns the *relevance of place and territory* in urban development, demonstrated in the serial reproduction of inner London industrial districts and working-class residential communities. Contemporary London represents an adaptation from the Chicago school of social ecology which stressed the internal patterns of human settlement and activity shaped by durable class structure. Assuredly place-identity retains its value as an analytical construct. But increasingly 'place and territory' in the global city form the basis of identity appropriation, marketing and social upgrading, rather than as secure habitus for diverse social groups.

Beyond the serial upgrading of sites within the established zones of central and inner London, spatial change for London in this century has included the extension of the economy and new housing deeper into outer London, with gentrification extending its reach within neighbourhoods located in Zones

226 Brexit london

3 and 4 of the Underground. Highly remunerated London professionals seek pièd-a-terres in the amenity-rich regions of southern England as well as in France, Spain and Italy: forming a high-amenity socio-cultural realm for these favoured cohorts.

A fourth defining spatial feature of the London story concerns the idea of the *'extended metropolitan region'* (EMR), introduced as a model of Asian peri-urban development by Terry McGee and his colleagues, influenced by Jean Gottmann's original concept of megalopolis (1961). At the larger European scale, London functions as a strategic node of a network of cities including Paris, Berlin, Amsterdam, Brussels and Barcelona, characterised by complementarity and cooperation as well as competition. The nature, governance and operating characteristics of London's prominent position within this European urban network will be increasingly disrupted by the effects of Brexit.

Another critical aspect of urban-regional development for Britain's capital includes a form of *city-centred innovation zones and corridors*, exemplified by the Boston-Route 128 region in the eastern seaboard, and the San Francisco-Bay Area region in northern California. In the London-centred South East region, a thickening of inter-institutional linkages between London, the University of Oxford and the University of Cambridge form a leading, global-scale innovation zone: the latter universities conventionally ranked 1 and 2 respectively among the world's leading post-secondary institutions, and with Oxford ranked first in the world for medicine and Cambridge first in science. London, Oxford and Cambridge comprise the apex sites of what is known colloquially as Britain's 'Golden Triangle' of propulsive sectors, knowledge-intensive institutions and human capital, connected by collaboration and partnerships.

While the key institutions in the Golden Triangle include universities and related research centres and sites, government is actively engaged in fostering development in this critical sector, following an entrepreneurial model of statecraft. As a measure of the policy distance travelled for the extended development region centred on London and including adjacent counties I recall my survey based research from the 1980s (Hutton 1991), which disclosed generally restrictive attitudes of local authorities in parts of London and more especially at the county and district level in the South East Region concerning the merits of development, even in areas designated for growth in regional planning guidelines.

Overall London sustains its privileged position within an expanding roster of globalising cities within the Atlantic realm, the growth economies of East Asia and the cities of the Global South. A recent article in the *Financial Times* reported on surveys which disclosed that foreign banks and asset management companies had maintained staffing levels in London since the Brexit referendum of 2016, and in some cases had actually increased personnel (*Financial Times* 12 December 2020). At some point though the structural weakness of the British economy, deeply flawed national polity and multi-scalar social divisions at national and community levels are likely to undermine London's status within the network of global cities.

Recasting spatial effects of redevelopment: The value of the London experience

As this volume demonstrates, processes of globalisation, capital relayering and cultural practices reproduce the spaces of the city, encompassing an ever-increasing range of places and territories. The scalar implications of change include London's positionality within transnational networks of production and innovation; the central role of London within Britain's 'Golden Triangle' of higher-level innovation and knowledge creation; the enlarged zonal structure of deep capital relayering and redevelopment within the GLA extending well beyond the central area; and of course the (re)production of space at the district, community and site levels within the metropolis: the thematic framing for Chapters 4–9.

What follows is concise elucidation of key effects of these processes and experiences at three scalar levels: strategic platforms for the global city; the recasting of urban territories for new enterprise and enhanced value capture by market players; and a selective representation of the iconography of sites in London.

Platform effects: The basis of global city power and projection

Platforms represent the most strategic feature of contemporary cities, comprising in many cases new capital projects as well as more established ensembles of specialised economic activity and employment. Platforms typically incorporate clusters, industrial districts, networks and development corridors described earlier. A defining feature of platforms in this century is the meshing of place-situated assets with digital technologies that facilitate the operation of advanced production regimes, and high-density content transmission across space, while enabling firms and workers to enjoy the qualities of place (histories, imageries, built environment) which attract talent.

Global cities, such as London, have experienced (and in many cases actively fostered) the expansion of these strategic clusters (or platforms) through new concentrations of financial and human capital, often stimulated by regeneration programs entailing collaboration between the local state and market interests. They entail a substantial spatial expansion of the urban economy, even in mature cities like London, and represent a hallmark of twenty-first-century urbanism and the quest for competitive advantage. Relatedly, platforms are important for infrastructural investments, specialised industries and high-value labour and complementary consumption and amenity features.

In London twenty-first-century platforms include notably *finance*, with the City of London and Canary Wharf international finance and business complex at its heart, and with specialised institutions and firms in other Central

228 Brexit london

London locations; *innovation and higher education*, anchored by Imperial College, the London School of Economics and Political Science (LSE), King's College London and University College and many other tertiary institutions, accompanied by the rise of inner north-east London as zone of culture-infused technological innovation; *culture and creativity*, with key institutions such as the Victoria & Albert Museum, British Museum and the National Gallery to name just a few of the key sites that support London (together with Washington DC) as the leading world city in terms of cultural institutions, and supported by the myriad sites of creative enterprise described in Chapter 6, exemplified by the Camden Cultural Corridor; and finally London's numerous *megaprojects*, including the massive Vauxhall-Nine-Elms-Battersea megaproject centred in Wandsworth, and the subject of Chapter 4 of this volume.

Each of these is underpinned by London's strategic *transportation and communications infrastructure and systems*, a platform which connects London to the world (notably through Heathrow as well as Gatwick and City Airport), and facilitates movement and circulation within the metropolis via the Underground and surface transportation modes.

These platforms enable a substantial projection of the British capital's economic power and cultural influence. Accordingly they feature centrally in city promotional campaigns, and typically incorporate cultural allusion, a mix of historical and contemporary signifiers and messaging concerning the power of civic values and virtues.

The concentrations of capital associated with London's platforms and ancillary land use also shape a disruptive upgrading and dislocation tendency. To illustrate: Edward Jones and Jessica Ferm of the Bartlett School UCL (2015) have written about the pressures on industrial land exerted by intensive (and extensive) capital relayering throughout the metropolis. They conclude that the continuing loss of industrial land—key to start-ups, small-scale production, warehousing and distribution—is shaped by real estate speculation rather than by deindustrialisation.

Territorial effects: The appropriation of local histories and cultural signifiers

London demonstrates that 'territory' comprises a salient feature of growth and change in the contemporary metropolis. In London territory is situated as a spatial construct comprising cultural/historical signifiers, social actors and institutions, with Bloomsbury in southern Camden an example, and within mature cities often entailing functional overlays and succession as new and emergent actors and groups with access to capital and other resources effectively 'make over' historic areas for high-margin activities.

Territories in the city typically comprise mixed-use areas incorporating specialised economic activity, neighbourhoods and consumption and retail activity. Mature metropolitan cities such as New York, Milan, Shanghai and of course London abound in multiple 'territories'—the stuff of rich urban narratives, fiction, social identity formation and opportunistic 'capture' by property interests and other market players, and the materièl for penetrating city ethnography in the tradition of Rebecca Solnit, Richard Lloyd and Michael Indergaard

In the London case as in other prominent examples 'territory' is deployed as an overlay, palimpsest or aggregation of sites embodying historic meaning as part of area-based marketing and local area regeneration. In London as we have seen Clerkenwell is one of the most resonant territories, incorporating longitudinal traditions of high-value artisanal production as well as (in the late modern era) upscale housing and consumption. As Alan Ainsworth's rich monograph disclosed, an array of new actors—artists, developers and residents—has reproduced elements of Clerkenwell's historic industrial specialisation, although an insistent upgrading experience since the 1990s has pushed lower-income residents and workers further out.

Territory in London can also apply to areas of convivial interaction by social groups within existing areas of specialised production, including the vast array of bars and cafes in areas like those extending from the Silicon Roundabout through the Shoreditch Triangle and more recently into the edges of Tower Hamlets.

In my usage 'territory' in London can also incorporate multiple land uses and capital projects, including the megaprojects arrayed along the River from Chelsea to the Victoria Dock. While some critics might depict these megaprojects as sites of inert sunk capital, their cultural significance, including the appropriation of historical signifiers, imparts a particular territorial aspect to London's development trajectory.

The nomenclature in London includes the 'new frontier' territory of the Victoria Dock in East London which now encompasses multiple capital projects downstream from Canary Wharf. As Tim Butler observes, the distance of these new projects and neighbourhoods from the established zones and territories along the Thames corridor, and the lower density of human presence and interaction, lends an almost suburban quality to what is intended as a new urban community on the eastern fringe of the global city.

Sites and place-based effects: Contested meanings in the global city

A third register of space in contemporary London is situated in more localised areas of the metropolis, at the level of place, sites or even individual buildings and memorials. With this sharper spatial delineation of place comes more acute debate (and conflict) over meaning and experience, both in historical and contemporary settings. In this final section I identify a compact list of examples, designed to acknowledge the exceptional range of sites and places of memory and commemoration in London.

The Borough Market in Bermondsey was one of London's principal wholesale markets, and now presents as an especially lively site of social consumption and interaction, with foodstuffs sourced locally and elsewhere in Britain and from

230 Brexit london

continental suppliers. The Borough Market also represents a resonant exemplar of the cultural makeover of postindustrial Southwark in the heritage spaces of the twenty-first-century city. But the Borough Market also represents as site of terrorism and violence, represented by the deaths from the attack of 3 June 2017. A public commemoration of the event takes a form of the planting of an olive tree in (adjacent) Southwark Cathedral: a symbol of healing and reconciliation through history.

The striking Kindertransport memorial site proximate to Liverpool Street Station commemorates the effort of the British government to bring Jewish children to Britain in the period before the Second World War. While the site evokes the memory of murdered Jewish children enacted far from Britain's capital, executed in the Nazi extermination camps in Germany and Poland, the wrenching imageries underscore the quality of the Holocaust as unique in the modern era.

Trafalgar Square, certainly one of London's most resonant places for locals and visitors alike, presents one of the most exigent sites of cultural contestation. According to one's vantage point, Trafalgar Square represents a totalising naval victory over the combined Franco-Spanish fleet under Villeneuve, effectively putting an end to the prospect of invasion; a fitting testimonial to British admiral Horatio Nelson, who defeated the combined fleet at the cost of his life (and those of many more); or an unwelcome and outdated celebration of empire (and colonial oppression and slavery), no longer fit for purpose. But for many foreign tourists Trafalgar Square is mostly a convenient meeting place and access point to the National Gallery, St Martins in the Fields and other attractions.

Processes of development, decline and regeneration have the power to reshape the imagery of sites and spaces. These include London's rail stations, defining infrastructural features of Victorian London. King's Cross and St Pancras typify a transformation from the postwar deterioration of public infrastructure and dereliction of adjacent sites, to the decidedly twenty-first-century spectacle of high-level consumption and visual experience and including restoration of the iconic station sheds.

Fifteen years ago I wrote in a journal article (2006) a story about the symbolic transformation of Farringdon Station, on the City Fringe and just north of the Smithfield Market. In the 1860s Farringdon comprised part of the world's first metropolitan underground railway, the Metropolitan Line. By the time of writing (Hutton 2006), the station area had experienced decline, with the scope of cultural regeneration shifting further north in Clerkenwell and more particularly to the Shoreditch area centred on Hoxton Square.

Cowcross Street, across from Farringdon Station, was characterised by low-margin retail operations and informal consumption outlets. But as part of the regeneration effects associated with the massive public investment in the Elizabeth Line, revitalisation of the Farringdon Station recaptures some of the iconicity of its foundation a century and a half ago, and with Cowcross Street now replete with upscale dining, specialty coffee houses and boutique shops, as I recounted in the previous chapter.

My final example of the meanings and imaginaries of site in the remaking of London concerns Renzo Piano's Shard tower. As I described in Chapter 4, the Shard was constructed on the former site of a perfectly serviceable if aesthetically prosaic PriceWaterhouse office tower. A Labour Council approved the demolition of this 1980s-era commercial tower to enable the construction of the Shard—on the face of it a decision inconsistent with the sustainability values of a progressive executive body.

The meshing of design elegance and architectural projection in Piano's Shard contrasts with the faux-whimsy of the City's newer office towers across the River, evoked in the popular terminology of 'the Gherkin' and 'the Cheesegrater', among other descriptors: effectively critiqued by Maria Kaika. These buildings comprise newer landscape features of the City—a genuine power centre within the heart of London's most resonant business complex, while the Shard accommodates a more promiscuous mix of uses.

The Shard is Europe's most striking new building in the twenty-first century, visible across much of London, and increasingly identified as a landscape icon of the British capital. Continuing with the emphasis of this volume on space and relational geographies of urban development, the Shard also occupies a strategic location, representing a lynchpin of London's twenty-first-century development program: directly across from the City of London; with the Borough Market, Tate Modern and the Globe Theatre to the west; Bermondsey's culture-led regeneration program to the south; and with the vast expanse of high-margin megaprojects extending mile after mile to the east. The Shard represents both empirically and symbolically a strategic shift in the centre of gravity of metropolitan London eastward in the present era: capturing an important moment in the life of this great city.

Coda: Spaces of plague and pandemic, London 1665 and 2020

London is the capital and site of central government within a unitary state, situated on an island 30 kilometres from the continental mass of Europe. This insularity hasn't however constrained London's remarkable development shaped by external trade relations, financial flows, immigration and cultural exchange with states and societies in Europe and the rest of the world since the sixteenth century. As I have discussed in this book, London represents one of the most instructive cases of the relational geographies of urban development across realms which include finance, culture and higher education among other fields of human enterprise.

As the record of urbanisation and city network flows attests, there are risks as well as benefits associated with the engagement of cities and urban societies in international trade, travel and migration. Risks can include infection across a spectrum of disease vectors, associated with travellers from foreign states. These risks can be managed—but in most cases not eliminated—by border controls, commitment to community health and effective vaccination and other management protocols.

The postwar experience in Britain and Western Europe in managing health risks associated with communicable disease infections is in many ways impressive, owing to innovation in public health, research on the causes of illness and management of disease vectors, dietary improvement and much enhanced information sharing: including, notably, the exchange of knowledge, joint research and common funding of health and disease prevention within the programs and partnership protocols of the European Union.

Britain's experience with the novel coronavirus pandemic of 2020, in terms of infections and deaths, has been among the worst of western European nations, owing in part to population densities, including poor standards of public housing in some cases and to widespread disregard of distancing protocols on the part of segments of the population. It's fair to say as well that the British government has not ranked among the European leaders in the management of the pandemic, with policies and programs fragmented and with uneven delivery of treatment. The British government can take some credit for the vastly improved Covid situation in the spring of 2021. But arguably the most important factors have included the leadership of key researchers at the University of Oxford and partner institutions engaged in vaccine development, the remarkable commitment of health professionals and myriad volunteers both to (initially) targeted group and mass vaccination, and the strength of public health systems and institutions in Britain overall.

I conclude *Millennial Metropolis* with a concise reflection on selected aspects of comparison between the experience of the 2020 pandemic in London with the effects of the bubonic plague of 1665. As in the other fields of human activity and social interaction described in this volume, the quality of 'space' is relevant to an assessment of causality and effects.

Of course the contrasts between the British capital's experience of pandemic situated in two periods three and half centuries apart are to say the least salient. But there are at the same time striking commonalities to observe, notably in respect to effects of population densities, neighbourhood characteristics and socioeconomic conditions on rates of contagion and death.

For the London plague experience of 1665, source materials include the rich quotidian entries on how the disease played out among different areas of the city recorded by two great diarists, John Evelyn and Samuel Pepys. Evelyn and Pepys each held important positions in the state, and each has also left a record of how London's population experienced the ravages of a plague far more deadly than 2020's Covid pandemic.

As context we can observe some defining commonalities between the two eras under consideration here. Political upheaval was common to each period. In 1665 as in 2020 Britain was engaged in disputes with continental powers, especially with regard to control of trade. That said the current (October 2020) struggle between the UK and the European Union over terms of the former's departure, although certainly immensely consequential, is pitched at a lower register than the series of wars with Holland, and rivalry with other European nations, particularly France, that characterised the seventeenth century.

The plague of 1665 produced a death toll in London many times higher than that of the Covid pandemic of 2020. In September of 1665 deaths from the plague reached 10,000 a week in London, reaching a total of 70,000 by the end of October (Saunders 1970: 75). The crisis of disease was compounded by the strain of related effects, including starvation associated with a diminished supply of food from the countryside. John Evelyn was responsible for the welfare of prisoners of war, and appealed to the King for assistance in preventing the starvation of 5,000 French prisoners. To Evelyn's credit he and his fellow commissioners of the Navy provided £6,000 for emergency relief of starving prisoners out of their own pockets, although then as now even the most generous of benefactors cannot make up for deficiencies in the provision of state resources in times of crisis.

In the plague year of 1665, place of residence in London and its region was highly consequential in terms of risk from infection, and here Samuel Pepys' residency in the City was advantageous. As Robert Latham, co-editor (with William Matthews) of the definitive University of California Press version of the Pepys' diaries has written, environmental conditions were critical to relative rates of infection and fatality. In this regard higher propensity for fatality from plague infection experienced by residents of London outside the City:

> … was not because the bacillus was less virulent or toxic inside the walls than in the liberties or out-parishes but simply because in the City people were thinner on the ground. The City buildings consisted largely of noblemen's big houses surrounded by quite spacious gardens, or else of business premises only partially occupied at night. People slept, and died, for the most part, in appallingly crowded tenements outside the walls – in Holborn or Clerkenwell, in Cripplegate or Bishopsgate, or along the Fleet River, or along the Whitechapel Road and the Ratcliffe Highway. Further dense populations with corresponding high mortality were to be found along the river in Wapping and Rotherhithe, or across it in Southwark or Bermondsey, or in the then almost isolated villages of Stepney or Shoreditch.
>
> *Latham 335, in Latham and Matthews (1983)*

Rates (and fatal consequences) of infection owing to Covid in London in 2020 are far lower than those recorded for the plague year of 1665. And unlike the plague of 1665 or the other epochal pandemics of the middle ages, an effective vaccine has been approved and has been widely administered. But what is common to the plague years and the 2020-21 Covid pandemic is the correlation between spatial density and housing quality, and rates of infection and death.

The geography and social composition of impacts of Covid in London are in some respects complicated. The second phase of infections (autumn 2020) included populations resident in some of the more prosperous areas of the metropolis, notably Richmond, while higher infections among younger Londoners have been recorded, owing in part to an apparent disregard for social distancing protocols among many.

But the initial phase of infection in London in late February and March and summer of 2020 followed to a degree the vector of the 1665 plague, with markedly higher rates of infection impacting denser communities associated with lower socioeconomic status. Data gleaned from the Greater London Authority's City Intelligence Unit disclosed comparatively high rates of infection in areas of East London, such as Newham and Tower Hamlets.

The experience of other areas of London disclosed the interdependency between Covid infection rates and socioeconomic status, including the quality of housing as well as income. Seriously afflicted boroughs included Brent in north-west London, which incorporates areas of congestion in housing estates which represent fertile grounds for infection.

To illustrate, The Guardian's journalist Aamna Modhin reported on the situation in low-income areas of Brent, including a large Somali population, where she observes that high Covid infection rates and fatal outcomes were associated with 'housing pressures, in-work poverty and racial inequalities' contributing to the death of 36 residents of the Church End neighbourhood (https://theguardian.com/worlds/audio/2020/jul03/how-one-neighbourhood-brent-london-lost-36-residents-to-covid-19).

A clear implication of this succinct comparison of the 1665 plague and 2020 pandemic experience in London is that 'space matters': not just in the siting of technological innovation, new enterprise formation and labour market change, nor only in the social use of space for human interaction and convivial relations, but also for the conferring of privilege and security for favoured groups, and higher risks from disease for less affluent communities.

APPENDIX: METHODS AND SOURCES

My study of London takes the form of a theoretically informed reading of the value of space in assessing both formative processes and consequences of change in London, realised through instructive case studies, lively storylines and comparison with reference cases situated in other globalising cities.

I'm interested in a framing of change shaped in large measure by shifts in governance, statecraft and policy values in the metropolis. London represents a particularly instructive case of urbanism owing to its extended record of continuity across key fields of human enterprise, notably in governance, finance and trade. But London's development trajectory has been increasingly subject to deeper and more pervasive disrupture initiated by Margaret Thatcher's neoliberal agenda, privileging market principles and competition in the reproduction of space.

Successive governments, both Conservative and New Labour, have tinkered with neoliberalism at the margins, exemplified by support for local regeneration programs which offer opportunity for diverse social groups. But the record shows that state agencies in London have largely accepted the basic tenets of neoliberalism as essential to competing for capital, skilled labour and market share in key sectors.

Points of departure: A half-century of restructuring in London

London in 1970 presented as a mature metropolitan city, its distinctive spatiality layered by the institutions of the state and the financial and corporate office complex situated with the City. Privileged space in Central London was allocated to religious assembly, major cultural and educational institutions and social elites.

236 Methods and sources

Fifty years later much of this template remains, although an insistent upgrading experience has remixed the social composition of many neighbourhoods in favour of new elites.

The London of 1970 also featured extensive industrial lands structured by specialised divisions of production labour along with affiliated working-class neighbourhoods within much of East London, including the large Ford auto plant at Dagenham established in 1931. Other significant representations of industry were situated in north-western areas such as Kilburn and Wembley, and districts south of the Thames including Battersea and Rotherhithe. The Port of London, for more than a century the world's largest, experienced a terminal decline over the 1970s. But a vast infrastructure of docks, warehouses and cranes remained as derelict residuals of London's maritime trading vocation, recurrently deployed in place-remaking by the market and state agencies in this century.

Over the half-century since the collapse of London's specialised industrial sectors, working-class communities and riverside port functions, these extensive territories have experienced serial processes of capital relayering, upgrading and dislocation. At the heart of this social transformation is the growth of a 'new middle class' of professionals and managers, documented notably by Chris Hamnett, Tim Butler and Loretta Lees among other scholars, and in this century incorporating cohorts of 'supergentrifiers' possessing far deeper financial resources than their predecessors. Relatedly the interests of these privileged cohorts have been accommodated by an assemblage of banks, property companies, media and compliant state actors which underpin a potent 'territorial financialization' program in the British capital.

Acknowledging diversity in urban change

As crucial as these elite professionals and upper-tier wealth holders have been to the production of a new social order in the British capital, spatial change in twenty-first-century London incorporates a more diverse template of actors and agencies that cannot be accommodated within a single theoretical construct.

The spatial outcomes of these actors and processes exhibit variegation across the metropolis, situated within an expanded zonal template which extends beyond the established districts of the central and inner city. I present a spatial model of London's propulsive sectors and capital projects at the conclusion of Chapter 3.

Change in London is situated at different scalar registers, and over different temporal spans. The City of London, notably, exhibits a durable primacy within the capital's financial and business districts, but has undergone significant institutional change and the introduction of new architectural imageries since the 1990s. Shoreditch, a compact industrial district in East London, has been subject to internal change in product and labour specialisation since the 1980s; has extended its influence on adjacent spaces and sites through spillover effects and policy mimicry; and was the subject of an effort by (then Prime Minister) David Cameron to link this iconic cultural zone to the development possibilities of the former 2012 Olympic site of Stratford.

As a means of drawing out the implications of spatial change in London, my research approach incorporates ideas framed within the following conceptual ensembles, comprising a rich field of empirically informed theory. 'Empirical' I interpret as incorporating reference to data, notably in assessing the social impacts of occupational and employment change; and also a careful reading of landscape features and the built environment as cues to interpreting change, notably in the serial upgrading of sites and spaces.

Industrial urbanism and place-based innovation in the reshaping of territory

Referential theory for Millennial Metropolis includes in the first instance a variant of industrial urbanism, with the industrial district in its successive iterations comprising a centrepiece of change in the city, articulated in Allen Scott's (1988) depiction of complex interdependency between divisions of labour and urban form. Technological disruption, changes in the enterprise model of many high-margin businesses and the ongoing restructuring of labour markets have produced widespread dislocations of firms and labour *in situ*.

The effects of the Covid-19 pandemic on labour markets, work practices and the structures of enterprise will take the form of permanent change, as well as the massive contractions of firms and employment in 2020 and continuing well into 2021. But the London case demonstrates that the contemporary city is still relevant to understanding practices of industrial innovation, network formation and an affiliated reproduction of space.

A reading of contemporary urban ethnography, within which specific spaces of the city are deployed as territories of innovative business and labour practices, contributes to a deeper understanding of the recasting of social actors in the spaces of the global city. Michael Indergaard's (2004) saga of creative innovation, disruption and eventual decline and collapse in New York's Manhattan mid-town area offers a lively entrée into the complex interactions of capital, culture, enterprise and social actors in the spaces of the global city. London offers a particularly instructive case study of these processes, comprising the complexity of actor-space interaction in districts situated within the City Fringe in particular.

Over the past two decades lively storylines of industrial innovation and succession in London comprise a critical literature addressing territorial upgrading. As former sites of specialised industries and craft labour have given way to cohorts of artists, creative professionals and specialists in digital technology innovation and marketing, London scholars have established an influential discourse on what this means in terms of changing industrial governance, and local policy values and practices, in the global city.

238 Methods and sources

The appropriation of cultural imaginaries in capital projects and urban change

Assessment of change framed within key spatial constructs in Millennial Metropolis is informed by a careful reading of the critical policy studies literature on London and other globalising cities. Here I situate change in state values and city planning practices which produce space-shaping outcomes for which London stands as a pioneering site. I direct a critical gaze on change within London's diverse places and territories, while bringing in reference to other globalising cities where such a practice is instructive.

The most spectacular landscape expressions of neoliberalism in London takes the form of a sequence of capital projects arrayed along the Thames, starting with the Canary Wharf global financial complex and more recent high-density residential projects situated downstream toward the estuary. A westward extension of the phenomenon includes the massive Nine Elms-Vauxhall-Battersea project, Chelsea Barracks and the Chelsea Reach-Lots Road projects.

Canary Wharf involved a comprehensive remaking of the Isle of Dogs, and a political discounting of industry and its social cohorts and cultural values. Michael Heseltine's agenda implied an obliteration of the complex industrial experiences and social narratives of the site, *en route* to the remaking of a new London dominated by finance, property development and business services. But successive megaprojects in London are reproduced by means of an appropriation of cultural memory and social values in the formation of project imaginaries and marketing programs.

This is obviously the case in the spectacular remaking of the former Bankside Power Station as the Tate Modern, widely regarded as a subject of remarkable community affinity with historical building form for Londoners. But 'culture' as embodied within building design, social memory and artfully curated industrial objects situated within the precincts of the Battersea and Lots Road power stations present impactful signifiers of the larger transition from sites allocated for public purpose to market orientation.

Postcolonialism, vernacular cultures and 'everyday globalization'

The axial narrative of change in London since 1970 is formed by the principal ensembles of financialisation, capital relayering, megaprojects and new industrial districts, together with affiliated social cohorts endowed with high levels of cultural and financial capital. But accounts of these principal drivers of change in London can leave out much of what contributes to London's cosmopolitan quality.

Accordingly in Millennial Metropolis I identify (and locate through case studies) space for the remarkably diverse social actors and groups which animate London's storyline at the neighbourhood scale. Rather than comprising social residuals of larger processes, I position neighbourhoods of immigrant populations, working class families and mixed spaces of actors as crucial to the vitality of London as global metropolis.

Here I'm influenced by writers who have contributed to rich discourses of postcolonialism, the lives of working class Londoners striving to find a place in the revalorised city and the role of affordable places of consumption and experience in animating the 'spaces in between' the sites of capital relayering and social upgrading in London.

Millennial Metropolitan: Methodology and sources

The heart of my research base for this book comprises a program of key informant interviews, field work observation and cultural mapping conducted over the first two decades of this century, coupled with an engagement with the rich literature on London's experience of urbanism. This period takes in a deepening of capital layering in established communities, the serial upgrading of labour within important districts and attendant dislocation of enterprise and workers associated both with industrial innovation and the effects of financialisation shaped by market actors.

Key informant interviews and information sharing

I have benefitted from the generosity of colleagues prepared to share specialised knowledge of critical aspects of London's growth and change. These included during my time of graduate research in the School of Geography, University of Oxford, faculty members (notably my initial research supervisor Professor Jean Gottmann FBA, Professor J. W. House and Dr Ian Scargill). I benefitted from the generosity of London scholars situated at other universities, notably Peter Hall (Reading and Birkbeck College), Peter Damesick (Birkbeck College), Gerald Manners (University College London), Peter Daniels (Liverpool) and Derek Diamond (LSE).

As graduate student I learned much about the operation of state agencies in London's important experience of growth and change from government officials, including, as examples, meetings with Dr John Palmer (Department of the Environment) and Paul Gibson, GLC office policy specialist in the research program for the Greater London Development Plan, described in Chapter 2 of this volume. More recently I have benefitted from discussions and information exchange with professional staff at the Greater London Authority, as well as planning and regeneration specialists in boroughs such as Islington, Hackney and Southwark.

I acknowledge the generosity of colleagues prepared to share their deep knowledge of London with me, including Andy Thornley, Graeme Evans, Jo Foord, Andrew Harris, Chris Hamnett, Andy Pratt, Tony Travers and Max Nathan; and, at UBC, David Ley and Trevor Barnes.

240 Methods and sources

Comparative study of urban change: Locating London in relational geographies of development

Millennial Metropolis is designed as case study of internal change in London as shaped by agencies and actors over time. I have endeavoured to enrich the case study with reference to development experiences in other globalising cities where I believe they demonstrate general processes or, in other respects, more contingent effects.

These include commentary on comparative policy values shaping decisions for urban megaprojects, including aspects of public risk and alignment with globalisation aspirations (Chapter 4); complex factors which produce cultural space in the global city (Chapter 5); reference to cities experimenting with programs for enhancing conviviality (Chapter 6); acknowledgement of cities aspiring to foster innovation districts as part of larger economic development programs and the pursuit of competitive advantage (Chapter 7); the rise of new cohorts of super-gentrifiers imbued with sufficiently deep financial resources to compete with (and in many cases displace) earlier groups, observed in apex cities like London, New York and San Francisco but also smaller and more peripheral cities like Vancouver (Chapter 8); and finally a hopeful commentary on what I characterise as spaces of vernacular culture and mixed social actors in the heart of the global city, informed by Timothy Shortell's lively study of 'everyday globalization' in Brooklyn and Paris (Chapter 9).

Primary sources, governance discourses and policy reports

Shifts in the values, content and presentation formats of governance documents pertaining to London and its larger region represent important markers of change over the last half-century. The broader pathway is marked by a transition from policy reports imbued with values of political economy, mostly concerned with instruments of development control, and in many cases prepared over protracted time periods; to the contemporary modality of policies for London shaped by a more positivist and even ebullient spirit.

For my graduate research on London in the 1970s and 1980s, I compiled an extensive dossier of policy and planning documents. These include reports prepared by the Standing Conference on London and South East Regional Planning (SCLSERP), and the South East Economic Planning Council (SEEPC) from 1972 to 1976, as well as county structure plans and district plans.

The narrative of the pre-Thatcher era was shaped by an appreciation of the values of influential postwar plans intended to achieve the 'containment' of London within its regional setting as starting point, addressed in Peter Hall's magisterial 2-volume set, *The Containment of Urban England* (George Allen & Unwin: PEP: 1973).

This spirit of containment as policy value and planning rubric was embodied in documents associated with the *Greater London Development Plan*, and plans prepared by individual London boroughs, and regional bodies noted earlier, and cited in Chapter 2 of this study. The underlying assumption for most of these reports was that 'growth' in London (economic activity and employment as well as population) represented a first-order policy problem, and must be stringently managed, both in the interests of preserving the integrity of the Metropolitan Green Belt, and also in restraining the regional imbalance of employment growth described in Doreen Massey's *Spatial Divisions of Labour* (1984).

The underlying premise of this extensive lineage of policy reports, shaped by concern about the externalities of metropolitan dominance within the region and nation as a whole, was compromised by the economic crises of London shaped by the decline of the Port of London and collapse of manufacturing. These governance values and policy precepts were comprehensively repudiated in Margaret Thatcher's neoliberal agenda starting with her Conservative Party's election in 1979 and in the ensuing period of London's waxing primacy within Britain.

I have accessed data on aspects of London's growth and change derived from a range of sources, including census data generated and published by the government of Britain and its ministerial agencies; data published by London government, including the Greater London Council (to 1986); the Greater London Authority; and individual London boroughs. I carried this program of documentary access forward to the era of the Greater London Authority since its establishment in 2000.

In particular I acknowledge the value of data sets and analytical reports published by the GLA's Intelligence and Analysis Unit. These provide a wealth of data and analytical commentary on London's experience of growth and change. As example I cite an instructive report published by the GLA (authored by Mike Hope) titled *The Economic Impact of Brexit on London* (GLA October 2019), and including forecasts of lower business investment, and declining output and productivity, ensuing from Brexit in almost any form.

At the local level each of the London boroughs publishes a broad range of reports including economic, labour and employment, housing and cultural trends and issues. The London boroughs also publish policy reports and progress assessments on regeneration programs at the local and sub-regional levels. These are published in many cases in draft or provisional form for public consultation purposes, and include statements of community objectives and project deliverables.

As an example I cite here the *Vauxhall-Nine Elms-Battersea Opportunity Area Planning Framework* (November 2009), a complex document setting out issues and opportunities associated with a large area of riverine Wandsworth, which I describe in Chapter 4 of this book. The 174 pp. document includes detailed descriptions of the site, including environmental conditions, the built environment, histories and heritage features and opportunities for public engagement in the detailed planning to follow.

As a measure of change in the publication values of policy documents published by state agencies on London over the last 50 years, those dating from the

242 Methods and sources

early period tended to embody austere design qualities and a restrained tonality, oriented towards policy professionals; while many of the publications on London's development possibilities in this century feature a more upbeat messaging and higher production values.

Part of this difference in presentation values over time is of course associated with technological innovation in publishing. But the more positivist tone of contemporary publications on London's development aligns with the exuberantly promotional quality of urbanism in the neoliberal era.

Program of London field studies

An extended program of area studies in London comprised a key element of my research on change at the district and community levels, conducted within a sample of globalising cities over the last three decades. Over the past quarter-century my program has evolved from an initial investigation of change in industries and firms as measures of change in land use, to a more rounded program encompassing territorial histories, characteristics of the built environment and aspects of both interdependency between production sectors and consumption activity in the precincts of the city.

My field research program for London includes site visits over a quarter-century, starting with initial visits in 1995 and 1998, and then in this century 2000, 2002, 2003, 2004, 2006 (two visits), 2009, 2012, 2014, 2016, 2017, 2018 and 2019.

My larger program of field work in globalising cities, which have assisted in identifying relational aspects of development in London, include initial field observations in San Francisco's South of Market Area (SOMA) on the growth of creative industries and labour was undertaken in 1993 with return visits for interviews and cultural mapping in 1995 and 1997. A program of key informant interviews, mapping and field photography was undertaken over the following decade: 2000, 2001, 2003, 2005 and 2007.

A visiting scholar tenure at the National University of Singapore in 1999 (and with follow-up visits in 2000, 2003 and 2006) afforded me opportunity to study the 'new economy' in Chinatown, with a focus on cultural mapping in the Telok Ayer district. In this century Trevor Barnes and I extended parallel field work on cultural mapping and interviews in Seattle and Vancouver. I have reported on these observations in two earlier books for Routledge, *The New Economy of the Inner City* (2008), and *Cities and the Cultural Economy* (2016). These projects afforded opportunity to study aspects of contingency within the general experiences of growth and change in cities situated within the ambit of globalisation.

Field experiences have proven useful in identifying larger patterns of district-level change in London, as well as in recognising aspects of contingency associated with urban system dynamics, scale effects, governance and planning models. From the late 1990s through to 2019 I have conducted serial field study and field work in three instructive London districts: Clerkenwell on the City Fringe, Shoreditch (Hackney and borderlands of Tower Hamlets) and Bermondsey (Southwark).

More recently I conducted field study of signifying capital projects in riverine London: the Nine Elms-Vauxhall-Battersea projects in Lambeth and Wandsworth; the early stages of the Chelsea Barracks project directly across the river; and the Lots Road Power Station and Chelsea Reach projects on the Chelsea–Fulham border. Downriver from Canary Wharf I also conducted field work in the Victoria Dock, in connection with the 'new-build gentrification' theme developed by Tim Butler, and which comprises an important eastward trajectory of capital projects in London.

BIBLIOGRAPHY

Abbas, A. (1997) *Hong Kong and the Politics of Disappearance*. Minneapolis, MN: University of Minnesota Press.

Agnew, J. (2005) *Hegemony: The New Shape of Global Power*. Philadelphia: Temple University Press.

Ainsworth, A. (2010) *Clerkenwell: Change and Renewal*. London: Alan Ainsworth and Obliqueimage.com.

Allen, J. (2000) 'On Georg Simmel: Proximity, Distance and Movement', Chapter 2, in: M. Crang, and N. Thrift (eds.), *Thinking Space* (pp. 54–70). London: Routledge.

Amin, A. and Graham, S. (1997) 'The Ordinary City', *Transactions of the Institute of British Geographers* 22: 411–429.

Amin, A. and Thrift, N. (eds.) (2004) *The Blackwell Cultural Economy Reader*. Oxford and Malden, MA: Blackwell.

Appadurai, A. (1996) *Modernity at Large: Cultural Dimensions of Globalization*. Minneapolis, MN: University of Minnesota Press.

Arena, R. and Quéré, M. (eds.) (2004) *The Economics of Alfred Marshall: Revisiting Marshall's Legacy*. New York: Palgrave MacMillan.

Arrighi, G. (1994) *The Long Twentieth Century*. London: Verso.

Arrighi, G. (2007) *Adam Smith in Beijing: Lineages of the Twenty-First Century*. London: Verso.

Atkins, M. and Sinclair, I. (1999) *Liquid City: Second Expanded Edition*. London: Reaktion Books.

Atkinson, R., Parker, S., and Burrows, R. (2017) 'Elite Formation, Power and Space in Contemporary London', *Theory, Culture and Society* 33: 1–25.

Barnes, T. (2001) 'Retheorizing Economic Geography: From the Quantitative Revolution to the "Cultural Turn"', *Annals of the Association of American Geographers* 91: 546–565.

Barnes, T. and Hutton, T. (2016) *Dynamics of Economic Change in Metro Vancouver: Networked Economies and Globalizing Urban Regions*. Metro Vancouver: Vancouver.

Bauman, Z. (1973) *Culture as Praxis*. London, Thousand Oaks CA, and New Delhi: Sage Publications.

Bauman, Z. (1998) *Globalisation: The Human Consequences*. New York: Columbia University Press.

Becattini, G. (1989) 'Sectors and/or Districts: Some Remarks on the Conceptual Foundations of Industrial Economics', in: E. Goodman, J. Bamford, and P. Saynor (eds.), *Small Firms and Industrial Districts in Italy* (pp. 125–135). New York and London: Routledge.

Bell, D. (1972) 'The Cultural Contradictions of Capitalism', *Journal of Aesthetic Education* 6: 11–38.

Bell, D. (1973) *The Coming of Postindustrial Society: A Venture in Social Forecasting.* New York: Basic Books.

Bell, David. and Jayne, M. (eds.) (2004) 'Conceptualizing the City of Quarters', in: *City of Quarters: Urban Villages in the Contemporary city* (pp. 1–12). Aldershot, UK: Ashgate.

Bianchini, F. and Ghilardi, L. (2004) 'The Culture of Neighbourhoods: A European Perspective', in: D. Bell, and M. Jayne (eds.), *City of Quarters: Urban Villages in the Contemporary City* (pp. 237–248). Aldershot, UK: Ashgate.

Bishop, R., Philips, J. and Yeo, W. W. (2004) *Beyond Description. Singapore, Space,, Historicity.* London and New York: Routledge.

Bluestone, B. and Harrison, B. (1982) *The Deindustrialization of America: Plant Closing, Community Abandonment, and the Dismantling of Basic Industry.* New York: Basic Books.

Bolton, T. (2017) *Camden Town: Dreams of Another London.* London: The British Library.

Boschma, R. (2005) 'Proximity and Innovation: A Critical Assessment', *Regional Studies* 39: 61–74.

Bourdieu, P. (1991) *Language and Symbolic Power.* Cambridge, MA: Harvard University Press.

Boyer, C. (2000) 'Is a Finance-Led Growth Regime a Viable Alternate to Fordism? A Preliminary Analysis', *Economy and Society* 29: 111–145.

Branzanti, C. (2014) 'Creative Clusters and District Economies: Toward a Taxonomy to Interpret the Phenomenon', European Planning Studies 23:1–18.

Braudel, F. (1966[1972]) *The Mediterranean and the Mediterranean World in the Age of Philip II*, I and II. New York, San Francisco and London: Harper & Row.

Brenner, N., MacLeod, G., Jessop, B., and Jones, M. (eds.) (2003) *State-Space: A Reader.* Oxford: Blackwell.

Bryson, J. and Daniels, P. W. (eds.) (2007) *The Handbook of Service Industries.* Cheltenham, UK: Edward Elgar.

BPSDC (2014) *Community Charter.* London: Battersea Power Station Development Corporation.

Butler, T. (2007) 'Re-Urbanizing London Docklands: Gentrification, Suburbanization, or New Urbanism?', *International Journal of Urban and Regional Research* 31: 759–781.

Butler, T., Hamnett, C., and Ramsden, M. (2013) 'Gentrification, Education and Exclusionary Displacement in East London', *International Journal of Urban and Regional Research* 37: 556–575.

Butler, T. and Lees, L. (2006) 'Supergentrification in Barnsbury, London: Globalization and Gentrifying Global Elites at the Neighbourhood Level', *Transactions of the Institute of British Geographers* 31: 467–487.

Campos, N. and Corricelli, F. (2017) 'How EEC Membership Drove Margaret Thatcher's Reforms', *VOX CEPR Policy Portal.* https://voxceu.org/article/how-eec-mmembership-drove-margaret-thatcher's-reforms [accessed 15 May 2020].

Capello, R. and Faggian, A. (2005) 'Collective Learning and Relational Capital in Local Innovation Processes', *Regional Studies* 39: 75–87.

Castells, M. (1996) *The Rise of the Network Society, the Information Age: Economy, Society and Culture Volume I.* Cambridge, MA and Oxford: Blackwell.

246 Bibliography

Catungal, J. P., Leslie, D., and Hii, Y. (2009) 'Geographies of Displacement in the Creative City: The Case of Liberty Village, Toronto', *Urban Studies* 46: 1095–1114.

Chang, T. C. (1995) 'The "Expatriatisation" of Holland Village', Chapter 7, in: B. S. A. Yeoh, and L. Kong (eds.), *Portraits of Places: History, Community and Identity in Singapore* (pp. 140–157). Singapore: Times Editions.

Chang, T. C. and Yeoh, B. H. (1999) 'New Asia—Singapore: Communicating Local Cultures Through Global tourism', *Geoforum* 30: 101–115.

Cherry, B. and Pevsner, N. (1983) *The Buildings of England – London 2: South*. London: Penguin Books, and New Haven, CT: Yale University Press.

Cherry, B. and Pevsner, N. (1998). *The Buildings of London – London 4: North*. London: Penguin Books, and New Haven, CT: Yale University Press.

Christopherson, S. (2003) 'Review of J. Cowie and J. Heathcott (Ed) Beyond the Ruins: The Meanings of Deindustrialization' (Ithaca NY: Cornell University Press: 2003), *Journal of the American Planning Association* 70: 487.

Chua, B. H. (1995) 'That Imagined Space: Nostalgia for *kampungs*', Chapter 11, in: B. S. A. Yeoh, and L. Kong (eds.), *Portraits of Places: History, Community and Identity in Singapore* (pp. 222–241). Singapore: Times Editions.

City Fringe Partnership (2005) *Analysing the Creative Sector in the City Fringe*. London: A Report by TBR Economics.

City of San Francisco (1997) *Multimedia in San Francisco*. San Francisco: City of San Francisco Planning Department.

Clark, T. N. and Lloyd, R. (2004) *The City as an Entertainment Machine*. Amsterdam and Boston: Elsevier.

Clark, T. N. and Lipset, S. M. (1991) 'Are Social Classes Dying?', *International Sociology* 6: 399–410.

Clark, T. N. and Lipset, S. M. (2001) *The Breakdown of Class Politics*. Baltimore, MD: The John Hopkins University Press.

Conzen, M., Dahman, N. M., and Schuble, T. J. (2006) *At Home Downtown: The residential transformation of Chicago's new global era core: 1981–2005'*. Report of the Chicago Committee on Geographical Studies, University of Chicago.

Cowie, J., Heathcott, J., and Bluestone, B. (2003) *Beyond the Ruins: The Meanings of Deindustrialization*. Ithaca, NY: Cornell University Press.

Crang, M. (2000) 'Relics, Places and Unwritten Geographies in the Work of Michel de Certeau (1925–1986)', Chapter 6, in: M. Crang, and N. Thrift (eds.), *Thinking Space* (pp. 136–153). London: Routledge.

Crang, M. and Thrift, N. (eds.) (2000) *Thinking Space*. London: Routledge.

Crinson, M. (ed.) (2005) *Urban Memory: History and Amnesia in the Modern City*. Abingdon (Oxon) and New York: Routledge.

Cuthbert, A. R. (2007) 'Urban Design: Requiem for an Era—Review and Critique of the Past 50 Years', *Urban Design International* 12: 177–223.

Currid-Halkett, E. (2008) *The Warhol Economy: How Fashion, Art and Music Drive New York City*. Princeton, NJ: Princeton University Press.

Cybriwsky, R. (2011) *Roppongi Crossing: The Demise of a Tokyo Nightclub District and the Reshaping of a Global City*. Athens, GA: University of Georgia Press.

Dell'Agnese, E. and Anzoise, V. (2011) 'Milan, the Unthinking Metropolis', *International Planning Studies* 16: 217–235.

Daniels, P. W. (1978) 'Confusing LOB', *Town and Country Planning* 46: 414–418.

Daniels, P. W., Ho, K. C., and Hutton, T. A. (eds.) (2012) *New Economic Spaces in Asian Cities: from Industrial Restructuring to the Cultural Turn*. Abingdon and New York: Routledge.

Darwin, J. (2012) *Unfinished Empire: The Global Expansion of Britain*. London: Penguin Books.

Davidson, M. and Lees, L. (2005) 'New-Build "Gentrification" and London's Riverside Renaissance', *Environment and Planning A* 37: 1165–1190.

Dearden, L. (2017) 'London Market Demolition Triggers UN Investigation into Area's gentrification', The Independent online: 27 October: https://www.independent.co.uk/news/uk/home-news/london-gentrification-investigation-seven-sisters-market-demolition-pueblito-paisa.

de Certeau, M. (1984) *The Practice of Everyday Life*. Berkeley: University of California Press.

Dennis, R. (2008) 'Babylonian Flats in Victorian and Edwardian London', *The London Journal* 33: 233–247.

Department for Culture, Media and Sport (2001) *Creative Industries Mapping Document*. London: DCMS and the Creative Industries Task Force.

Dorling, D. (2013) *The 32 Stops: The Central Line*. London: Penguin.

Drake, G. (2003) '"This Place Gives Me Space": Place and Creativity in the Creative Industries', *Geoforum* 34: 511–524.

Duménil, E. and Levy, D. (2005) 'Costs and Benefits of Neoliberalism: a Class Analysis', in: G. A. Epstein (ed.), *Financialization and the World Economy* (pp. 17–45). Cheltenham, UK: Edward Elgar.

Edensor, T. (2005) *Industrial Ruins: Space, Aesthetics and Materiality*. London: Bloomsbury Academic.

Edensor, T. (2005) 'The Ghosts of Industrial Ruins: Ordering and Disordering Memory in Excessive Space', *Environment and Planning D (Society and Space)* 23: 829–849.

Edensor, T., Leslie, D., Millington, S., and Rantisi, N. (eds.) (2012) *Spaces of Vernacular Creativity: Rethinking the Cultural Economy*. London and New York: Routledge.

Elden, S. (2013) *The Birth of Territory*. Chicago and London: University of Chicago Press.

Ensor, R. (1936 [1968]) *England 1870–1914*. Oxford: at the Clarendon Press.

Epstein, G. A. (ed.) (2005) *Financialization and the World Economy*. Cheltenham, UK and Northampton, MA: Edward Elgar.

Evans, G. (2001) *Cultural Planning: An Urban Renaissance?* London: Routledge.

Evans, G. (2003) 'Hard Branding the Cultural City: From Prado to Prada', *International Journal of Urban and Regional Research* 27: 417–440.

Evans, G. (2004) 'Cultural Industry Quarters: From pre-Industrial to Post-Industrial Production', in: D. Bell and M. Jayne (eds.), *City of Quarters: Urban Villages in the Contemporary City*. Aldershot: Ashgate.

Evans, G. (2009) 'Creative Cities, Creative Spaces and Urban Policy', *Urban Studies* 46: 1003–1040.

Fainstein, S. (2011) *The Just City*. Ithaca, NY: Cornell University Press.

Fanzini, D., Casoni, G., and Bergamini, I. (eds.) (2014) *Enhancement of Cultural Heritage and Development of the Territory*. Milan: Politecnico di Milano.

Ferm, J. and Jones, E. (2015) *London's Industrial Land: Cause for Concern?* London: The Bartlett School, University College London.

Ferm, J., Jones, E., and Edwards, M. (2017) *Revealing Local Economies in London: Methodological Challenges, Future Directions*. London: The Bartlett School, University College London.

Flyvberg, B. (2003) *Megaprojects and Risk: An Anatomy of Ambition*. Cambridge: Cambridge University Press.

Foord, J. (2013) 'The New Boomtown? Creative City to Tech City in East London', *Cities* 33: 51–60.

248 Bibliography

Foot, J. (2001) *Milan Since the Miracle: City, Culture and Identity*. Oxford and New York: Berg.

Fox, C. (ed.) (1992) *London – World City 1800–1840*. London: Yale University Press, and New Haven, CN: The Museum of London.

French, S., Leyshon, A., and Wainwright, T. (2011) 'Financing Space, Spacing Financialization', *Progress in Human Geography* 35: 798–819.

Friedmann, J. (1986) 'The World City Hypothesis', *Development and Change* 17: 69–83.

Friedmann, J. and Wolff, K. (1982) 'World City Formation: An Agenda for Research and Action', *International Journal of Urban and Regional Research* 6: 309–344.

Fröbel, F., Heinrichs, J., and Kreye, O. (1980) *The New International Division of Labour*. Cambridge: Cambridge University Press.

Giddens, A. (1991) *Modernity and Self-Identity: Self and Society in the Late Modern Age*. Cambridge: Polity.

Gill, R. and Pratt, A. C. (2008) 'In the Social Factory? Immaterial Labour, Precariousness and Cultural Work', *Theory, Culture & Society* 25: 1–30.

Glaeser, E. L., Kolko, J., and Saiz, A. (2001) 'Consumer City', *Journal of Economic Geography* 1: 27–50.

Glass, R. (1963) Introduction to London: Aspects of Change. London: Centre for Urban Studies. (Reprinted in R. Glass (ed.) (1989) *Clichés of Urban Doom*, Oxford: Blackwell.

Global and World Cities Group (2014) 'Global Civil Society Hyperlink Perspective: Exploring the on-line networks of international NGOs', GaWC Bulletin No. 439 www.lboro.ac.uk/gawc/rb/rb/439.html.

Glucksberg, L. (2016) 'A View from the Top: Unpacking Capital Flows and Foreign Investment in Prime London', *City*, 20: 238–255.

Goto, K. (2012) 'Craft and Creativity: New Economic Spaces in Kyoto', Chapter 6, in: P. W. Daniels, K. Ho, and T. A. Hutton (eds.), *New Economic Spaces in Asian Cities: From Industrial Restructuring to the Cultural Turn* (pp. 87–101). Abingdon and New York: Routledge.

Gottmann, J. (1961) *Megalopolis: The Urbanized Northeastern Seaboard of the United States*. New York: Twentieth Century Fund.

Gottmann, J. (1973) *The Significance of Territory*. Charlottesville, VA: University Press of Virginia.

Gottmann, J. (1990) 'Orbits: The Ancient Mediterranean Tradition of the Urban Networks', in: J. Gottmann (ed.), *Since Megalopolis: The Urban Writings of Jean Gottmann*, Baltimore, MD: The Johns Hopkins University Press.

Gottmann, J. (2005) *La Politique des etats et leur géographie*. Paris: CTHS.

Grabher, G. (2001) 'Ecologies of Creativity: The Village, the Group, and the Heterarchic Organisation of the British Advertising Industry', *Environment and Planning A* 33: 351–374.

Graham, S. and Marvin, S. (2001) *Splintering Urbanism: Networked Infrastructures, Technological Mobilities, and the Urban Condition*. London: Routledge.

Granger, R. (2015) 'A Tale of Two Cities: Spatial Change through Rent Gap and Spatial Fixes, Shifting Capital Flows and the Gentrination of the Global Semi-Periphery', Presentation to the Annual Meeting of the Association of American Geographers: Chicago.

Gravier, J. F. (1947) *Paris et le Desert Français*. Paris: Flammarion.

Greater London Authority (2002) *Preliminary Development Report*. London: GLA and The Office of the Mayor.

Greater London Authority (2012a) *World Cities Cultural Report*. London: GLA and The Office of the Mayor.

Greater London Authority (2012b) *Vauxhall Nine Elms Battersea Opportunity Area Planning Framework*. GLA: March.

Greater London Authority/L. Togni (2015) *The Creative Industries in London*. London: GLAEconomics.

Greater London Authority/M. Hope (2019) *The Economic Impact of Brexit on London*. London: GLAEconomics.

Greater London Authority (2020) Improved Decision-making for Infrastructure Resilience: Final Report: 3 October, ARUP and UCL.

Grodach, C. and Loukaitou-Sideris, A. (2007) 'Cultural Development Strategies and Urban Revitalization: A Survey of U.S. Cities', *International Journal of Cultural Policy* 13: 349–370.

Grodach, C. and Silver, D. (eds.) (2013) *The Politics of Urban Cultural Planning*. Abingdon (Oxon) and New York: Routledge.

Hall, P. G. (1962) *The Industries of London Since 1861*. London: Hutchinson.

Hall, P. G. (1966) *The World Cities*. London: Weidenfeld & Nicolson.

Hall, P. G. (1998) *Cities in Civilisation*. London: Weidenfeld & Nicolson.

Hall, P. G. (2000) 'Creative Cities and Economic Development', *Urban Studies* 37: 639–651.

Hall, P. G. (2006) *'The Polycentric City', PowerPoint Presentation*. London: The Bartlett School, University College London.

Hall, P. G., Thomas, R., Gracey, H., and Drewett, R. (1973) *The Containment of Urban England, Vol. 2, The Planning System: Objectives, Operations, Impacts*. London: George Allen & Unwin and Sage Publications.

Hall, R. K. (1972) 'The Movement of Offices from Central London', *Regional Studies* 6: 285–292.

Hamnett, C. (1991) 'The Blind Men and the Elephant: Toward a Theory of Gentrification', *Transactions of the Institute of British Geographers* 16: 173–189.

Hamnett, C. (1994) 'Socio-Economic Change in London: Professionalisation Not Polarisation', *Built Environment* 20: 192–203.

Hamnett, C. (2003) *Unequal City: London in the Global Arena*. London: Routledge.

Hamnett, C. (2016) 'The Changing Class Composition of Big Cities: Professionalization, Proletarianization, or Polarization', presentation in the Green College 'The Next Urban Planet: Rethinking the City in Space and Time' series: University of British Columbia: 16 November.

Hamnett, C. and Randolph, B. (1984) 'The Role of Landlord Disinvestment in Housing Market Transformation: an Analysis of the Flat Break-up Market in Central London', Transactions of the Institute of British Geographers 9: 259–279.

Hamnett, C. and Randolph, B. (1988) 'Housing and Labour Market Change in London: A Longitudinal Analysis, 1971–1981', *Urban Studies* 25: 380–398.

Hamnett, C. and Whitelegg, A. (2007) 'Loft Conversion in London: From Industrial to Postindustrial Land Use', *Environment and Planning A* 39: 106–124.

Harding, A. and Blokland, T. (2014) *Urban Theory: A Critical Introduction to Power, Cities and Urbanism in the 21st Century*. London and Thousand Oaks, CA: Sage Publications.

Harris, A. (2008) 'From London to Mumbai and Back Again: Gentrification and Public Policy in Comparative Context', *Urban Studies* 45: 2407–2428.

Harris, A. (2012) 'Art and Gentrification: Pursuing the Urban Pastoral in Hoxton, London', *Transactions of the Institute of British Geographers* 37: 226–241.

Harris, A. (2013) 'Concrete Geographies: Assembling Global Mumbai through Transport Infrastructure', *City* 17: 343–360.

Harris, A. (2014) 'The Canary Wharf-isation of Shoreditch', *The Londonist* 4 February https://londonist.com/2014/02/the-canary-wharf-isation-of-shoreditch accessed 15 February 2020.

250 Bibliography

Harris, A. (2015) 'Vertical Urbanisms: Opening Up Geographies of the Three-Dimensional City', *Progress in Human Geography* 39: 601–620.

Harrison, B. (1992) 'Industrial Districts: Old Wine in New Bottles?', *Regional Studies* 26: 469–483.

Hartman, C. and Carnochan, S. (2002) *City for Sale: The Transformation of San Francisco.* Berkeley and Los Angeles: University of California Press.

Harvey, D. (1982) *The Limits to Capital.* Chicago: University of Chicago Press.

Harvey, D. (1989) 'From Managerialism to Entrepreneurialism: Transformation in Urban Governance in Late Capitalism', *Geografiska Annaler Series B-Human Geography* 88B: 145–158.

Harvey, D. (2001) Spaces of Capital: Toward a Critical Geography. New York and Abingdon: Routledge.

Harvey, D. (2003) *Paris: Capital of Modernity.* London and New York: Routledge.

Hebdidge, D. (1979) *Subculture: The Meaning of Style.* London: Methuen.

Helbrecht, I. (2004) 'Bare Geographies in Knowledge Societies—Creative Cities as Text and Piece of Art: Two Eyes, One Vision', *Built Environment* 30: 191–200.

Helleiner, E. (1994) *States and the Reemergence of Global Finance: From Bretton Woods to the 1990s.* Ithaca, NY: Cornell University Press.

Helleiner, E. (2010) 'A Bretton Woods Moment? The 2007–2008 Crisis and the Future of Global Finance', *International Affairs* 86: 619–636.

Hesmondalgh, D. (2007) *The Cultural Industries.* London: Sage.

High, S., MacKinnon, L., and Perchard, A. (eds.) (2018) *The Deindustrialized World: Confronting Ruination in Postindustrial Places.* Vancouver: UBC Press.

High Commissioner for Human Rights (2017) *News Release: London Market Closure Plan Threatens Dynamic Cultural Centre.* Geneva: United Nations, Office of the High Commissioner for Human Rights.

Hill, D. (2014) 'Regenerating Southwark: Urban Renewal Prompts Social Cleansing Fears': 7 October 2014. https://www./theguardian.com/society/2014/oct/07/southwark-london-regeneration-urban-renewal-social-cleaning-fears [accessed 20 October 2019].

Hirsch, A. (2017) 'Toppling Statues? Here's Why Nelson's Column Should Be Next', *The Guardian.* 22 August 2017. https://www.theguardian.com/commentisfree/2017/aug/22/toppling-statues-nelson's-column-should-be-next-slavery [accessed 15 August 2019].

Ho, K. C. (1994) 'Industrial Restructuring, the Singapore City-State, and the Regional Division of Labour', *Environment and Planning A* 26: 33–51.

Ho, K. C. (2009). The Neighbourhood in the Creative Economy: Policy, Practice and Space in Singapore', *Urban Studies* 46: 1187–1202.

Ho, K. C. and Hutton, T. A. (2012) 'The Cultural Economy in the Developmental City-State: A Comparison of the Chinatown and Little India Districts of Singapore', Chapter 14, in: P. W. Daniels, K. C. Ho, and T.A. Hutton, *New Economic Spaces in Asian Cities: From Industrial Restructuring to the Cultural Turn*, 220–236.

Ho, K. C. and Kim, W. B. (1997) 'Studying Culture and the Built Environment', Chapter 5, in: W. B. Kim, M. Douglass, S. C. Choe and K. C. Ho (eds.), *Culture and the City in East Asia* (pp. 209–226). Oxford: at the University Press.

Hopkins, K. (2016) 'In London You'll Want to Live Near These Premier League Stadiums', Mansion Global. https://mansionglobal.com/articles/in-london-you-ll-want-to-live-near-these-premier-stadius-28103 [accessed 14 June 2020].

Hracs, B. (2009) 'Beyond Bohemia: Geographies of Everyday Creativity for Musicians in Toronto', Chapter 6, in: T. Edensor, D. Leslie, S. Millington, and N. Rantisi (eds.), *Spaces of Vernacular Creativity: Rethinking the Cultural Economy* (pp. 75–88). London and New York: Routledge.

Bibliography 251

Hutton, T. A. (1991) *Local Office Policy in the Context of Strategic Planning and Structural Change: The case of the London-South East Region*. DPhil Dissertation, School of Geography, University of Oxford.

Hutton, T. A. (2000) 'Reconstructed Production Landscapes in the Postmodern City: Applied Design and Creative Services in the Metropolitan Core', *Urban Geography* 21: 285–317.

Hutton, T. A. (2004) 'The New Economy of the Inner City', *Cities* 21: 89–108.

Hutton, T. A. (2006) 'Spatiality, Built Form and Creative Industry Development in the Inner City', *Environment and Planning A* 38: 1819–1841.

Hutton, T. A. (2008 [2010]) *The New Economy of the Inner City: Restructuring, Regeneration and Dislocation in the Twenty-First-Century City*. London and New York: Routledge.

Hutton, T. A. (2016) *Cities and the Cultural Economy*. London and New York: Routledge Critical Introductions to Urbanism and the City.

Hutton, T. A. (2017) 'The Creative City', Chapter 11, in: J. R. Short (ed.), *A Research Agenda for Cities*, 137–150. Cheltenham, UK and Northampton, MA: Edward Elgar.

Hutton, T. A. (2019) 'The Cultural Economy in Cities', Chapter 12, in: Tim Schwanen and Ronald van Kempen (eds.), *Handbook of Urban Geography* (pp. 180–194). Cheltenham, UK and Northampton, MA:Edward Elgar.

Hutton, T. A. (2020) 'City on the Edge: Vancouver and Circuits of Capital, Control and Culture', Chapter 2, in: P. Gurstein, and A. Hutton (eds.), *City on the Edge* (pp. 47–74). Vancouver: UBC Press.

Illich, I. (1973) *Tools for Conviviality*. London: Calder & Boyars.

Imrie, R., Lees, L., and Raco, M. (2009) *Regenerating London: Governance, Sustainability and Community in a Global City*. London and New York: Routledge.

Indergaard, M. (2004) *Silicon Alley: The Rise and Fall of a New Media District*. London and New York: Routledge.

Indergaard, M. (2013) 'Beyond the Bubbles: Creative New York in Boom, Bust and the Long Run', *Cities* 33: 43–50.

Indergaard, M., Pratt, A. C., and Hutton, T. A. (eds.) (2013) Creative Cities after the Fall of Finance, special theme issue of *Cities*, 33: 1–9.

Jacob, D. (2010) 'Constructing the Creative Neighbourhood: Hopes and Limitations of Creative Policies in Berlin', *City, Culture and Society* 1: 193–198.

Jacobs, J. M. (1996) *Edge of Empire: Postcolonialism and the City*. London and New York: Routledge.

Jarvis, H. and Pratt, A. C. (2006) 'Bringing It All Back Home: The Extensification and 'Overflowing' of Work. The Case of San Francisco's New Media Households', *Geoforum* 37: 331–339.

Jeffries, S. (2016) 'An Invitation to the White House: We Go Inside Dagenham's Experimental Art Factory', *The Guardian*. 29 November.

Jencks, C. (ed.) (1992) *The Post-Modern Reader*. London: Academy Editions, and New York: St Martin's Press.

Jones, E. V. (2014) 'Governing Creative Economies in Inner East London', presentation to the annual meeting of the Association of American Geographers, Tampa: 8–12 April.

Kaika, M. (2010) 'Architecture and Crisis: Re-Inventing the Icon, Re-Imag(in)Ing London and Re-Branding the City', *Transactions of the Institute of British Geographers* NS35: 453–474.

Kaika, M. (2011) 'Autistic Architecture: The Fall of the Icon and the Rise of the Serial Object of Architecture', *Environment and Planning D* 29: 968–992.

Kaufman, N. (2009) *Place, Race and Story: Essays on the Past and Future of Historic Preservation*. Abingdon (Oxon) and New York: Routledge.

252 Bibliography

Kim, W. B., Douglass, M., Choe, S. C. and Ho, K. C. (eds.) (1997) *Culture and the City in East Asia*. Oxford: at the University Press.

King, A. D. (1976) *Colonial Urban Development: Culture, Social Power and Environment*. London and Boston: Routledge & Kegan.

King, A. (2016) *Writing the Global City: Globalization, Postcolonialism* and *the Urban* (Architext) Routledge; London and New York: London.

Knox, P. (1987) 'The Social Production of the Built Environment: Architects, Architecture and Post-Modernism', *Progress in Human Geography* 11: 354–377.

Knox, P. (2011) *Cities and Design*. Abingdon (Oxon) and New York: Routledge.

Kollowe, J.. (2015) 'Battersea Is Part of a Huge Building Project – But Not for Londoners', *The Guardian*. 14 February 2105. https://www.theguardian.com/business/2015/feb/battersea-nine-elms-propery-development-housing; accessed 4 March 2019.

Krätke, S. (2011) *The Creative Capital of Cities. Interactive Knowledge Creation and the Urbanization Economies of Innovation*. Malden, MA and London: Wiley-Blackwell.

Krätke, S. (2014) 'Cities in Contemporary Capitalism', *International Journal Urban and Regional Research* 38: 1660–1677.

Krätke, S. (2015) 'New Economies, New Spaces', Chapter 4, in: R. Paddison and T. Hutton (eds.), *Cities and Economic Change: Restructuring and Dislocation in the Global Metropolis*. London: Sage Publications.

Krätke, S. and Taylor, P. J. (2004) 'A World Geography of Media Centres', *European Planning Studies* 12: 459–477,

Krippner, G. R. (2005) 'The Financialization of the American Economy', *Socio-Economic Review* 3: 173–208.

Kynaston, D. (1994) *The City of London, Volume I: A World of Its Own*. London: Pimlico.

Landry, C. and Bianchini, F. (1995) *The Creative City*. London: Demos.

Lazeratti, L. (ed.) (2013) *Creative Industries and Innovation in Europe*. Abingdon (Oxon) and New York: Routledge.

Lambeth Borough Council (2018) *South London Innovation Council*. https://www.lambeth.gov.gov.uk/better/fairer/lambeth/project/south-london-innovation-corridor [accessed 20 February 2019].

Latham, R. C., 'The Plague' essay in *Volume X, Companion*, pp. 329–337, in Latham, R. C. and Matthews, W. (1985), *The Diary of Samuel Pepys*. London: Bell & Hyman, and Berkeley and Los Angeles: University of California Press.

Lee, D. J. (2011) 'Networks, Cultural Capital and Creative Labour in the British Independent Television Industry', *Media, Culture and Society* 33: 549–565.

Lees, L. (2003) 'Super-Gentrification: The Case of Brooklyn Heights, New York City', *Urban Studies* 40: 2487–2509.

Lees, L., Slater, T. and Wyly, E. (2008) *Gentrification*. London and New York: Routledge.

Lefèbvre, H. (1968) The Right to the City. New York: Verso.

Lefèbvre, H. (1974) *Production de l'espace* (Paris: Anthropos); published in English as *The Production of Space*, translated by D. Nicholson-Smith, 1991 (Oxford: Blackwell).

Lewis, R. (2004) 'Planned Districts in Chicago: Firms, Networks and Boundaries, 1900–1940', *Journal of Planning History* 3: 29–49.

Lewis, R. (2008) *Chicago Made: Factory Networking in the Industrial Metropolis*. Chicago: University of Chicago Press.

Ley, D. F. (1996) *The New Middle Class and the Remaking of the Central City*. Oxford: Geographical and Environmental Studies, OUP.

Ley, D. F. (2003) 'Artists, Aestheticization and the Field of Gentrification', *Urban Studies* 40: 2527–2544.

Ley, D. F. (2005) 'The Social Geography of the Service Economy in Global Cities', Chapter 4, in: P. W. Daniels, K. C. Ho, and T. A. Hutton (eds.), *Service Industries and Asia-Pacific Cities: New Development Trajectories* (pp. 77–92). Abingdon (Oxon) and New York: Routledge.

Livingstone, K. (2003) 'Commentary on the London Shard proposal', London Southeast 1 Community News online. [London-se1-uk/news/views/732] accessed 19 March 2020.

Lloyd, R. (2006) *Neo-Bohemia: Art and Commerce in the Postindustrial City.* New York and Abingdon, UK: Routledge.

Lloyd, R. and Clark, T. N. (2007) 'The City as Entertainment Machine', *Research in Urban Sociology* 6: 357–378.

London Higher (2017) 'Student Numbers in London' https://www/londonhigher.ac.uk/ceo-blog/student-numbers-in-london [accessed 23 October 2019].

LondonSE1 community paper (2017) 'Bermondsey Street's old Ticino Bakery: 7-room hotel approved'. https://www.london.se1.co.uk/news/view/9141 [accessed 6 September 2019].

McCann, P. (2016) *The UK Regional-National Economic Problem: Geography, Globalization and Governance.* London and New York: Routledge and the Regional Studies Association.

McDowell, L. (2012) 'Post-Crisis, Post-Fordism and Post-Gender? Youth Identities in an Era of Austerity', *Journal of Youth Studies* 15: 573–590.

McDowell, L. (2015) 'The Lives of Others: Body Work, the Production of Difference, and Labor Geographies', *Economic Geography* 91: 1–23.

McDowell, L. and Christopherson, S. (2009) 'Transforming Work: New Forms of Employment and their Regulation', *Cambridge Journal of Regions, Economy and Society* 2: 335–342.

Mckenzie, M. (2013) (Re)rembering the inner city: cultural production, reflexity and Vancouver's heritage areas. Unpub PhD dissertation. School of Community & Regional Planning, University of British Columbia: Vancouver.

Mckenzie, M. and Hutton, T. A. (2015) 'Culture-Led Regeneration in the Post-Industrial Built Environment: Complements and Contradictions in Victory Square, Vancouver', *Journal of Urban Design* 20: 8–27.

McRobbie, A. (1994) *Postmodernism and Popular Culture.* London and New York: Routledge.

Maier, J. (2015) *Rome Measured and Imagined: Early Modern Maps of the Eternal City.* Chicago and London: The University of Chicago Press.

Manthorpe, R. (2018) 'Airbnb Is Taking Over London', *Wired*: 24 October.

Markus, T. (1994) *Buildings and Power: Freedom and Control in the Origin of Modern Building Types.* London: Routledge.

Markusen, A. (1996) 'Sticky Places in Slippery Spaces: A Typology of Industrial Districts', *Economic Geography* 72: 293–313.

Markusen, A. (2006) 'Urban Development and the Politics of a Creative Class: Evidence from a Study of Artists', *Environment and Planning A* 38: 1921–1940.

Markusen, A. and Schrock, G. (2006) 'The Artistic Dividend: Urban Artistic Specialization and Economic Development Implications', *Urban Studies* 43: 1661–1686.

Martin, J. E. (1964) 'The Industrial Geography of Greater London', in: R. Clayton (ed.), *The Geography of Greater London* (pp. 111–142). London: George Philip and Son Limited.

Martins, J. (2015) 'The Extended Workplace in a Creative Cluster: Exploring Space(S) of Digital Work in Silicon Roundabout', *Journal of Urban Design:* 20: 125–145.

Massey, D. (1984) *Spatial Divisions of Labour: Social Structures and the Geography of Production.* London: MacMillan.

254 Bibliography

Massey, D. (2007) *World City*. London: Polity Press.

Massey, D. and Meegan, R. (1980) 'Industrial Restructuring versus the Cities', in: A. Evans and D. Eversley (eds.), *The Inner City: Employment and Industry*. London: Heinemann.

Massey, D. and Meegan, R. (1982) *The Anatomy of Job Loss*. London: Methuen.

Mayor of London (2012) *World Cities Cultural Report*. London: Mayor's Office.

Miles, M. (2007) *Cities and Culture*. Abingdon (Oxon) and New York: Routledge Introductions to Urbanism and the City.

Mitchell, D. (2003) *The Right to the City: Social Justice and the Fight for Public Space*. New York: Guilford Press.

Molotch, H. (1976) 'The City as a Growth Machine: Toward a Political Economy of Place', *American Journal of Sociology* 82: 309–332.

Molotch, H. (2002) 'Place in Product', *International Journal of Urban and Regional Research* 26: 665–688.

Molotch, H. and Treskon, M. (2009) 'Changing Art: SoHo, Chelsea and the Dynamic Geography of Galleries in New York City', *International Journal of Urban and Regional Research* 33 (2009) 517–41.

Moore, C. (2013) *Margaret Thatcher: The Authorized Biography. Volume 1: From Grantham to the Falklands*. London: Allen Lane, and New York: Alfred A. Knopf.

Moore, C. (2016) *Margaret Thatcher: The Authorized Biography. Volume II: At Her Zenith: In London, Washington and Moscow*. London: Allen Lane, and New York: Alfred A. Knopf.

Moore, C. (2019) *Margaret Thatcher: The Authorized Biography. Volume III: Herself Alone*. London: Allen Lane, and New York: Alfred A. Knopf.

Morris, W. (1890) *News from Nowhere*. London: Longmans.

Murray, G. (2004) 'Rethinking Neighbourhoods: From Urban Villages to Cultural Hubs', Chapter 12, in: D. Bell, and M. Jayne (eds.), *City of Quarters: Urban Villages in the Contemporary City* (pp. 191–205). Aldershot, UK: Eastgate.

Muscarà, L. (1998) 'Jean Gottmann's Atlantic "Transhumance" and the Development of His Spatial Theory", *Finisterra XXXII* 65: 159–172.

Nathan, M., Vandore, E. and Voss, G. (2018) 'Spatial Imaginaries and Tech Cities: Place-Branding East London's Digital Economy', *Journal of Economic Geography* 19: 409–432.

Newman, K. (2014) 'Commodifying Poverty: Gentrification and Consumption in Vancouver's Downtown Eastside', *Urban Geography* 35: 157–176.

Novy, J. and Colomb, C. (2013) 'Struggling for the Right to the (Creative) City in Berlin and Hamburg: New Urban Society, New "Spaces of Hope"? ', *International Journal of Urban and Regional Research* 37: 1816–1838.

Ocejo, R. E. (2010) 'What'll It Be? Cocktail Bartenders and the Redefinition of Services in the Creative Economy', *City, Culture and Society* 1: 179–184.

Olds, K. (2001) *Globalization and Urban Change: Capital, Culture, and Pacific Rim Megaprojects*. Oxford: Oxford University Press.

O'Loughlin, M. (2016) 'Snackistan London Restaurant Review', *The Guardian*. Online https://theguardian.com/lifeandstyle/2016/nov25/snackistan-London-restaurant-review-marina-oloughlin.

Paccoud, A. (2016) 'Private Rental-led Gentrification in England: Displacement, Commodification and Dispossession', Luxembourg Institute of Socio-Economic Research (LISER), WP Series 2015–2016: posted 2016.

Paddison, R. and Hutton, T. A. (eds.) (2015) *Cities and Economic Change*. London and Thousand Oaks, CA: Sage Publications.

Panayi, O. (2020) *Migrant City: A New History of London*. New Haven, CT: Yale University Press.

Pasotti, E. (2013) 'Brecht in Bogotá', in: C. Grodach, and D. Silver (eds.), *The Politics of Urban Cultural Policy* (pp. 42–53). Abingdon (Oxon) and New York: Routledge.

Peck, J. (2001) 'Neoliberalizing States: Thin Policies/Hard Outcomes', *Progress in Human Geography* 253: 445–455.

Peck, J. (2005) 'Struggling with the Creative Class', *International Journal of Urban and Regional Research* 29: 740–770.

Peck, J. (2013) 'Making Space for Labour', in: D. Featherstone, and J. Painter (eds.), *Spatial Politics: Essays for Doreen Massey* (pp. 99–114). New York and London: John Wiley & Sons.

Peck, J. (2015) 'Cities beyond Compare?', *Regional Studies* 49: 160–182.

Peck, J. and Theodore, N. (2015) *Fast Policy: Experimental Statecraft at the Threshold of Neoliberalism*. Minneapolis, MN: University of Minnesota Press.

Peters, W.. (1989) 'Terminal Illness', *Sunday Times*, 'New Society' Section F, p. 1.

Pike, A. (2013) 'Economic Geographies of Brands and Branding', *Progress in Human Geography* 33: 619–645.

Pile, S. (1996) 'The Strange Case of Western Cities: Occult Globalisation and the Making of Urban Modernity', *Urban Studies* 43: 305–318.

Popp, A. and Wilson, J. (2007) 'Life Cycles, Contingency and Agency: Growth, Development and Change in English Industrial Districts and Clusters', *Environment and Planning A* 39: 2975–2992.

Porter, R. (1994) *London: A Social History*. London: Hamish Hamilton.

Portoghesi, P. (1983) *Postmodern: The Architecture of the Post-Industrial Society*. Milan: Rizzoli.

Power, D. and Scott, A. J. (eds.) (2004) *Cultural Industries and the Production of Culture*. London and New York: Routledge.

Pratt, A. C. (1997) 'The Cultural Industries Production System: A Case Study of Employment Change in Britain', *Environment and Planning A* 29: 1953–1974.

Pratt, A. C. (2009) 'Urban Regeneration: From the Arts "Feel Good" Factor to the Cultural Economy: A Case Study of Hoxton, London', *Urban Studies* 46: 1041–1062.

Pratt, A. C. and Hutton, T. A. (2013) 'Reconceptualising the Relationship between the Cultural Economy and the City', *Cities* 33: 86–95.

Proll, A. (ed) (2011) Goodbye to London: Radical Art and Politics in the 1970s. Stuttgart: Hatje Cantz.

Punter, J. (2003) *The Vancouver Achievement: Urban Planning and Design*. Vancouver: UBC Press.

Raco, M. (2002) *Toward a London Plan: The 'Blue Ribbon' Network*. London: Greater London Authority.

Raco, M. (2005) 'A Step Change or a Step Back? The Thames Gateway and the Re-Birth of the Urban Development Corporations', *Local Economy* 20: 141–153.

Rantisi, N. (2002) 'The Competitive Foundations of Localized Learning and Innovation: The Case of Women's Garment Production in New York City', *Economic Geography* 78: 441–462.

Rantisi, N., Leslie, D., and Christopherson, S. (2006) 'Placing the Creative Economy: Scale, Politics and the Material', *Environment and Planning A* 38: 1789–1797.

Relph, E. (1981) *Rational Landscapes and Humanistic Geography*. New York: Barnes and Noble.

Relph, E. (1993) 'Modernity and the Reclamation of Place', in: D. Seamon (ed.), *Dwelling, Seeing and Designing: Toward a Phenomenology Ecology* (pp. 25–40). Albany, NY: SUNY Press.

Rizvi, F. (2011) 'Experiences of Cultural Diversity in the Context of an Emergent Transnationalism', *European Educational Research Journal* 10: 180–188.

256 Bibliography

Romein, A. and Trip, J. J. (2013) 'Notes of Discord: Urban Cultural Policy in the Confrontational City', Chapter 4, in: C. Grodach and D. Silver (pp. 54–68) *The Politics of Urban Cultural Policy*, Abingdon (Oxon) and New York: Routledge.

Robinson, J. (2006) *Ordinary Cities: Between Modernity and Development*. London: Routledge.

Rossi, I. and O'Higgins, E. (1980) 'The Development of Theories of Culture', Chapter 2, in: I. Rossi (ed.), *People in Culture* (pp. 31–78). New York: Praeger.

Ruskin, J. (1981[1853]) (abridged) *The Stones of Venice* (Introduction by J. Morris). Boston: Little, Brown and Company.

Sacco, P. L. and Tavano Blessi, G. (2009) 'The Social Viability of Culture-Led Urban Transformation Processes: Evidence from the Bicocca District, Milan', *Urban Studies* 46: 1115–1136.

Sack, R. (1983) 'Human Territoriality: A Theory', *Annals of the Association of American Geographers* 73: 55–74.

Sandercock, L. (1998) *Toward Cosmopolis: Planning for Multicultural Cities*. Chichester, UK: Wiley.

Sasajima, H. (2013) 'From Red Light District to Art District: Creative City Projects in Yokohama's Kogane-Cho Neighbourhood', *Cities* 33: 77–85.

Sassen, S. (2001[1991]) *The Global City: New York, London, Tokyo*. (Second Edition) Princeton, NJ: Princeton University Press.

Saunders, B. (1970) *John Evelyn and His Times*. Oxford: Pergamon Press.

Sawyer, M. (2013) 'What Is Financialization?', *International Journal of Political Economy* 42: 5–18.

Saxenian, A. (1991) 'The Origins and Dynamics of Production Networks in Silicon Valley', *Research Policy* 20: 423–437.

Scott, A. J. (1982) 'Locational Patterns and Dynamics of Industrial Activity in the Modern Metropolis: a Review Essay', *Urban Studies* 19: 111–142.

Scott, A. J. (1988) *Metropolis: From the Division of Labor to Urban Form*. Berkeley and Los Angeles: The University of California Press.

Scott, A. J. (1997) 'The Cultural Economy of the City', International Journal of Urban and Regional Research 21: 323–339.

Scott, A. J. (2000) *The Cultural Economy of Cities*. London: Sage.

Scott, A. J. (2008) *Social Economy of the Metropolis: Cognitive-Cultural Capitalism and the Global Resurgence of Cities*. Oxford: Oxford University Press.

Scott, A. J. (2011) *A World in Emergence: Cities and Regions in the 21st Century*. Cheltenham, UK: Edward Elgar.

Shabrina, Z., Arcaute, E., and Batty, M. (2019) 'Airbnb's Disruption of the Housing Structure in London'. Ithaca, NY: Cornell University. Computer Science≥ Computers and Society https://arxiv.org/pdf/1903.11205.pdf [accessed 15 June 2020].

Shaw, K. (2005) 'The Place of Alternative Culture and the Politics of Its Protection in Berlin, Amsterdam and Melbourne', *Planning Theory and Practice* 6: 151–170.

Shaw, K. (2013) 'Independent Creative Subcultures and Why They Matter', *International Journal of Cultural Policy* 19: 333–352.

Sheppard, F. (1998) *London: A History*. Oxford: Oxford University Press.

Shortell, T. (2016) *Everyday Globalization: A Spatial Semiotics of Immigrant Neighbourhoods in Brooklyn and Paris*. Abingdon (Oxon) and New York: Routledge Studies in Human Geography.

Smith, D. and Holt, L. (2007) 'Studentification and Apprentice Gentrifiers within Britain's Provincial Towns and Cities: Extending the Meaning of Gentrification', *Environment and Planning A* 39: 142–161.

Smith, M. P. (2001) *Transnational Urbanism: Locating Globalization*. Oxford: Blackwell.

Smith, N. (1979) 'Gentrification and Capital: Practice and Ideology in Society Hill', *Antipode 11*: 24–35.

Smith, N. (2002) 'New Globalism, New Urbanism: Gentrification as Global Urban strategy', *Antipode* 34: 427–450.

Soja, E. (2000) *Postmetropolis: Critical Studies of Cities and Regions*. Oxford: Blackwell.

Soja, E. (2010) *Seeking Spatial Justice*. Minneapolis, MN: University of Minnesota Press.

Solnit, R. and Schwartzenberg, S. (2000) *Hollow City: The Siege of San Francisco and the Crisis of American Urbanism*. London and New York: Verso.

Southwark Notes (2019) 'Student Housing as Best Performing Asset'. https://southwarknotes. wordpress.com/local-development-sites/studentification/ [accessed 22 October 2019).

Starr, K. (1996) *Endangered Dreams: the Great Depression in California*. New York and Oxford: Oxford University Press.

Stevenson, N. (2016) 'The Contribution of Community Events to Social Sustainability in Local Neighbourhoods', *Journal of Sustainable Tourism* 24: 990–1006.

Storper, M. (2013) *The Keys to the City: How Economics, Institutions, Social Interaction, and Politics Shape Development*. Princeton, NJ: Princeton University Press.

Storper, M. and Salais, R. (1997) *Worlds of Production: The Action Frameworks of the Economy*. Cambridge MA and London: Harvard University Press.

Surborg, B. (2006) 'Advanced Services, the New Economy and the Built Environment in Hanoi', *Cities* 23: 239–249.

Sutcliffe, A. (1973) *The Autumn of Central Paris: The Defeat of Town Planning, 1850–1970*. London: Edward Arnold. Studies in Urban History 1.

Swyngedouw, E. (2006) *The Post-Political City*. Manchester: School of Environment and Development, University of Manchester.

Theurillat, T. (2011) 'La Ville Negociée: entre financiarisaion et durabilité', *Géographie, Economie, Société* 13: 225–254.

Thornley, A. (ed.) (1992) *The Crisis of London*. London: Routledge.

Thornley, A. (2000) 'Dome Alone: London's Millennium Project and the Strategic Planning Deficit', *International Journal of Urban and Regional Research* 24: 689–699.

Thornley, A. (2012) 'The 2012 London Olympics: What Legacy?', *Journal of Policy Research in Tourism, Leisure and Events* 4: 206–210.

Torrance, M. (2009) 'The Rise of Global Infrastructural Markets through Relational Investing', *Economic Geography* 85: 75–97.

Travers, T. (2003) *The Politics of London: Governing an Ungovernable City*. London: Red Globe Press.

Travers, T. (2013) (quoted in)'London Project uses Risky Financial Model', *Financial Times* 8 April 2013.

Travers, T. (2015) *London's Boroughs at 50*. London: Biteback Publishers.

Turner, S. (2006) *Hanoi's Ancient Quarter: 'Traditional Traders' in a Rapidly Transforming City*. Montreal: McGill University: Department of Geography Research Paper.

Urry, J. (1995) *Consuming Places*. London and New York: Routledge.

Van den Berg, M. (2015) 'Imagineering the City', Chapter 10, in: R. Paddison, and T. A. Hutton (eds.), *Cities and Economic Change* (pp. 162–177). London and Thousand Oaks, CA: Sage Publications.

Van der Veer, P. (2002) 'Cosmopolitan Options', *Etnografica* VI: 15–26.

Van der Zwan, N. (2014) 'Making Sense of Financialization', *Socio-Economic Review* 12: 99–129.

Veblen, T. (1899) *The Theory of the Leisure Class: A Study of Institutions*. New York: The Macmillan Co.

258 Bibliography

Vertovic, S. and Cohen, R. 'Introduction: Conceiving Cosmopolitanism', Chapter 1, in: S. Vertovec and R. Cohen (eds.), *Conceiving Cosmopolitanism: Theory, Context and Practice* (pp. 1–22). Oxford: Oxford University Press.

Vijay, A. (2018) 'Dissipating the Political: Battersea Power Station and the Temporal Aesthetics of Development', *Open Cultural Studies* 2: 611–625.

Wachsmuth, D. and Weisler, A. (2018) 'Airbnb and the Rent Gap: Gentrification through the Sharing Economy', *Environment and Planning A* 50: 1147–1180.

Wainwright, O. (2019) 'The Mother of All Loft Conversions', Guardian on-line 16 September 2019. www.theguardian.com/artanddesign/2019/sep/16/tate-modern-best-architecture-21st-century-jacques-herzog [accessed 20 February 2020].

Wallerstein, I. (1974) 'The Rise and Future Demise of the World Capitalist System: Concepts for Comparative Analysis', *Comparative Studies in Society and History* 16: 387–415.

Ward, J. (2004) 'Berlin: The Virtual Global City', *Journal of Visual Culture* 3: 239–256.

Williams, C. C. (2015) 'Informal Economies', Chapter 7, in: R. Paddison, and T. A. Hutton (eds.), *Cities and Economic Change* (pp. 108–124). London and Thousand Oaks, CA: Sage Publications.

Willmott, P. and Young, M. (1957) *Family and Kinship in East London*. London: Pelican.

Wise, M. J. (1956) 'The Role of London in the Industrial Geography of Great Britain', *Geography* XLI.

Yeoh, B. and Kong, L. (1994) 'Reading Landscape Meanings: State Constructions and Lived Experience in Singapore's Chinatown', *Habitat International* 18: 17–35.

Yeoh, B. and Kong, L. (eds.) (1995) *Portraits of Places: History, Community and Identity in Singapore*. Singapore: Times Editions.

Yeung, H. and Lin, G. C. S. (2003) 'Theorizing Economic Geographies of Asia', *Economic Geography* 79: 107–128.

Yue, A. (2006) 'Cultural Governance and Creative Industries in Singapore', *International Journal of Cultural Policy* 12: 17–33.

Zhong, S. (2011) 'By Nature or Nurture: The Formation of New Economic Spaces in Shanghai', *Asian Geographer* 28: 33–49.

Zhong, S. (2012) 'New Economy Space, New Social Relations: M50 and Shanghai's New Art World in the Making', Chapter 11, in P.W. Daniels, K. C. Ho and T. A. Hutton (eds.), New Economic Spaces in Asian Cities: from Industrial Restructuring to the Cultural turn (pp. 166–183). London and New York: Routledge.

Zukin, S. (1989 [1982]) *Loft Living: Culture and Capital in Urban Change*. New Brunswick, NJ: Rutgers University Press.

Zukin, S. (1995) *The Cultures of Cities*. Oxford: Blackwell.

Zukin, S. (1998) 'Urban Lifestyles: Diversity and Standardisation in Spaces of Consumption', *Urban Studies* 35: 825–839.

INDEX

Italicized and **bold** pages refer to figures and tables respectively.

Abbey Road 102
Accra, mobilisation of space 224
Acton 198
agency 219–220
Ainsworth, A. 12, 229
Airbnb 172, 191–194
Albertopolis 128
Amanda Levete Architects 128
Amazon 144–146, 168
Amin, A. 8
Amsterdam 1, 4, 6, 41, 226
Andrews, J. 151
Antwerp 1, 41
Appadurai, A. 198
Arcaute, E. 192–193
architecture 4, 46, 53–55, *54*, 67, 70, 90, 94, 99, 193
Arnold Circus, Tower Hamlets 176, *177*, 199, 205–207, 216; Bandstand *207*
Arrighi, G. 47; *Adam Smith in Beijing* 221
Arsenal 12, 131, 132
arts, and gentrification 173–174
Asian financial crisis of 1997 67
Aston Villa 131
Athens 121, 122
Atkinson, R. 69

Babylonian flats 179
Ballard 144
Bangkok 6

Bank of England 42, 48
Bankside Power Station 73, 96, 222, 238
Barbican project 70
Barcelona 8, 92, 96, 226; convivial practices 120; mobilisation of space 224
Barnes, T. 144, 242
Battersea Power Station project 64, 66, 79, 80, 84, *85, 86, 91*
Batty, M. 192–193
Beijing 20
Belgravia, upgrading in 193
Bell, D. 7
Bell, The 71
Belltown 144
Berlin 6, 8, 96, 103, 226; mobilisation of space 224; social upgrading 175; symbolism of 129
Bermondsey 61, 73, 74, 79, 93, 103, 105, 118; arts, regeneration and upgrading, interdependencies of 109–111; Bermondsey Street: 2003–2018 111–113, 112; aesthetics of upgrading 113–117, *115, 116*; culture 113–117; property 113–117; Bermondsey Village Action Group 117; Borough Market 125–126, 229–231
Bermondsey Antique Market, Southwark 126, *127*
Bethnal Green 12, 13
Biden, J. 20
Big Lottery Fund 140
Billingsgate 125
Blackwall Reach project 75

260 Index

Blair, T. 34, 74, 219; New Labour program 35–37
Bloomberg Associates 151
Bloomberg, M. 151
Bloomsbury 96, 228
Bloomsbury Theatre 107
'Blue Ribbon' network concept 72
BNP-Paribas financial complex, Marylebone *51*
Boeing 144
Bond Street 113
Borough Council 79
Borough Market 125–126, 229–231
Boston 154
Bowling Green Lane 109
Boyer, C. 44
Braudel, F. 21
Brenner, N. 39; *State-Space* 20
Brent 61; SSE arena 134, *135*
Brexit 2, 4–5, 22, 38, 39, 43, 49, 68, 147, 218–234
Britannia Village 186–187; community site planning map *186*; housing form and heritage landscape *187*; project 75–76; recreational amenity sites *188*
British Library 107, 152
British Museum 98, 107, 127, 228
Brooklyn 17, 98, 167; everyday globalisation in 198, 240
Brussels 92, 226
Budapest 92
Burrows, R. 69
Butler, T. 7, 170, 172, 187–189, 209, 215, 229, 236, 243
Butler's Wharf 33, 73, 96, 180
Byzantium 121

Café Nero 203
Calatrava, S. 60
Calvert Street, Hackney 176–178, *177*
Cambridge University 61
Camden 23, 37, 59, 95, 111, 143, 151, 205; Camden Lock Market 101, 105, 137, *138*; Covent Garden 33, 107, 125, 181; cultural archipelago, morphology of 105–107, *106*; Cultural Corridor 228
Cameron, D. 36, 140, 163, 164, 166, 219, 237
Campos, N. 38
Canary Wharf 2, 12, 16, 35, 38, 42, 55–58, 60, 62, 64, 73–76, 79, 89, 133, 184, 219, 222, 225, 229, 238, 243; global financial cluster, corporate landscape of *56*; international financial

complex 49, 53, *57*, 63; transformations of space and territory 50
Cape Town 6
capital 195; cultural 69–70, 98, 117, 179; flows, localised 174–178; human 59, 145, 155, 167, 226, 227; projects 63–65, 183–189; appropriation of cultural imaginaries in 238; signifiers of 61–62; social 1, 2, 59, 97, 123, 137, 139, 140, 142, 168, 224
Capital Enterprise 152
capitalism 14; evolutionary 8; territorial 46–47, 62; twenty-first-century 3
Carter, J. 45
Cavendish Laboratory 61
Central London 203, 235; costs of rental housing in 191; gentrification in 171, 172, 188, 189, 194; iceberg houses in 179; land rents in 191; specialised health and medical institutions clusters in *148*; upgrading in 193
Central St Martins Art and Design College 71, *72*
Charles Dickens Museum 107
Chelsea 12, 37, 60, 82, 98, 132, 167; Chelsea Barracks project 73, 80, 88–89, *89*, 238, 243; Chelsea Bridge 80; Chelsea Creek 83; Chelsea Hospital 88; Chelsea Reach-Lots Road Station project 88, 238; Chelsea Reach project 88, 243; Chelsea Waterfront project 82–83; as frontier of megaproject development 88–89; gentrification in 178, 180; upgrading in 193
Cherry, B.: *London South* 111
Chicago 8, 14, 103, 144; Chicago School of Social Ecology 18, 225; industrialisation in 7; social upgrading 175; Wicker Park 15, 123, 160, 175–176
Chicago School 137
China 3
Choe, S. C.: *Culture and the City in East Asia* 93
Christopherson, S. 19
Cisco 159
City Corporation 48
City Fringe 12, 13, 26, 27, 35, 38, 60, 105, 110, 111, 145, 147, 159, 164, 187, 199, 207, 208, 230; Barbican project 70; social upgrading 176
City Island project 75, *76*
City of London Airport 74
City of London School for Girls 70
Clark, T. N. 123

Clerkenwell 49, 96, 103, 105, 111, 118, 199, 207–213, 217, 229, 242–243; 'Apulia' Pugliese restaurant, Cowcross Street *110*; Clerkenwell Parochial (Church of England) School 208–209; Clerkenwell Workshops 109, 212–213, *212*; Cowcross Street *214*, 230; culture and the reproduction of space in 107–109, *108*; Exmouth Market 209–210, *209*; Gazzano's Italian Deli and Cafe, Farringdon Road 210, *211*; Smithfield Market 213; Space and territory in 208–210; upgrading tendencies in 210–213
C40 network 6
coffee houses 52
Cohen, R. 124
community ethnographies 97–98
community practices 97–98
Concord Pacific Place megaproject 65
Conservative Party 241
convivial city 119–141; culture and 120; discourses and debates 122–124; experiences 127–129, *129*; expressions of 124; iconic institutions 127–129; markets 124–127, *126*, *127*; pandemic and political crises 140–141; as sites of consumption 121–124; society and 120; urbanism 120–121
Coopers Lybrand 77
Copenhagen 1, 6
Coram's Fields 107
Coricelli, F. 38
coronavirus (Covid) pandemic of 2020 22, 61, 67, 75, 120, 123, 141, 172, 192, 232–234, 237
Costa 203
Coubertin, Baron 122
Covent Garden 33, 107, 125, 181
Crang, M.: *Thinking Space* 14
creativity: social matrix of 97–98; territorial matrix of 102–103
Crinson, M.: *Urban Memory: History and Amnesia in the Modern City* 14
Crossrail high-speed system 149
Croydon 32
culture: aesthetic matrix of 101–102; and convivial practices 120; dislocation 104–105; and gentrification 173–174; cultural capital 69–70, 98, 117, 179; cultural encounter 195–197; cultural production districts/zones 102–103; cultural quarters 102; institutional matrix of 98–101; layering of 97–104;

of metropolis 95–97; of place 92–93; political—administrative matrix of 103; production of, social actors in 93–95; spaces of 92–93; succession 104–105; territorial ambit of 105–109; upgrading 104–105; of upgrading 117–118; vernacular 238–239

Dalston 105
Damesick, P. 239
Daniels, P. 239
Darwin, J. 41, 61, 90; *Unfinished Empire* 221
David Roberts Arts Foundation and Gallery 105, 107
Davidson, M. 87
Dearden, L. 182
Dear, M. 14
deindustrialisation 19
Delfina Studio Trust 111, 113
de Meuron, P. 81
Dennis, R. 23, 179
Department for Culture, Media and Sport (UK): Creative Industries Mapping Document 93
Department of the Environment 63, 71
Deva, S. 183
development trajectory 219–223
Dharavi, redevelopment of 169–170
Diamond, D. 239
Dickens, C. 13, 102, 122, 202, 207, 208
division of labour 2, 19, 25
Docklands 55–58, 62, 64, 90, 187, 188; Docklands Light Railway 75; Eastern Docklands redevelopment 66–67; megaprojects in 73–76, *74–77*; place-remaking in 77
Douglass, M.: *Culture and the City in East Asia* 93
Dubai 55
Duke of York 71
Dyson School of Design Engineering 150, *150*

East Asia 2, 3, 42, 55, 68, 74
East London 77, 94, 105, 132, 139, 236; capital relayering in 73–76, *74–77*; digital economy 163; gentrification in 173, 193–194; principal spatial development axis for 73; social upgrading in 178; Thameside, megaproject clusters in *74*
Edwardian era 49
EEC *see* European Economic Community (EEC)

262 Index

Eiffel Tower 90
Elden, S.: *Birth of Territory, The* 11
Emirates Stadium, Holloway 183
Empire Games (1934) 134
'empty house' syndrome 98
EMR *see* extended metropolitan
region (EMR)
Engels, F. 122
English National Ballet company 75
English Premier League (EPL) 131–132,
141, 171, 183
EPL *see* English Premier League (EPL)
Essen: as site of conviviality 122
ethnography: at the local scale 11–13
ethnospace 198
European Economic Community (EEC)
19, 29, 220
European Union 2, 22, 68; Britain's
membership in 9
Euston-St Pancras 107, 230
Evans, G. 102, 123
Evelyn, J. 232
everyday globalization 17, 98, 124, 180,
197–199, 202, *204*, 223, 238–240
evolutionary capitalism 8
ExCel Centre 75
Exhibition Road Quarter 128
extended metropolitan region
(EMR) 226

Fabian socialism 5
Facebook 159, 168
Fainstein, S.: *Just City, The* 6
Farrell, T. 88
Farringdon Station 230
Ferm, J. 195, 228
field studies program 242–243
financial crisis of 2008 48, 175
financialisation 43–47; London as site and
exemplar of 47–50; polarising effects of
45–46; territorial 90, 171, 236
Florence 4, 41, 92
Florence Nightingale Hospital 75
Flyvberg, B. 67
Foord, J. 143, 159–160
football 130–137, *134–136*; registers of
change 131–135
Football League First Division 131–132
Ford 34
Fordism 44
Foster, N. 58
Fox, C. 4
Frankfurt 6, 41
Fremont 144

French, S. 44
Friedmann, J. 96
Fulham 60, 79, 82, 83, 88, 131;
Fulham-Chelsea Reach 73

Gatwick 228
Geiger, K. 113
gentrification 5, 16, 17, 33, 35, 38, 82, 87,
98, 105, 136, 139, 143, 145, 169–195,
175, 199, 205, 225, 243; aesthetic
consumption-scape of *175*; in alpha
territories 178–179; arts and 173–174;
below the sight-lines 189–193; culture
and 173–174; industrial 172, 195; inner
city 179–183; in London 171–172; new
build 183–189; new industrial districts
174–178; regeneration as 181–183;
rent-gap 192
Gibson, P. 239
Glaeser, E. 123
GLA *see* Greater London Authority (GLA)
GLC *see* Greater London Council (GLC)
GLDP *see* Greater London Development
Plan (GLDP)
Global and World Cities (GaWC) project,
Loughborough University 14, 221
globalisation 3, 4, 8, 14, 16, 34, 44, 47–49,
53, 59, 67–69, 74, 90, 122, 123, 141,
156, 163, 193, 195, 198, 199, 202, 204,
215, 218– 221, 225, 227, 240, 242;
megaprojects and narratives of 65–66;
of space 63–65
globalising cities, platforms of
development in 2–3
Global South 2, 3, 14, 55, 226
global suburbanisms 8
Globe Theatre 113, 231
Glucksberg, L. 179
Golden Triangle 61, 226
Goldsmiths 151
Google 143, 159
Gottmann, J. 226, 239; 'Atlantic
transhumance' 11; *La Politique des Etats
at leur Géographie* 11; *Significance of
Territory, The* 11
governance 11, 36, 219–220; discourses
240–242; of Nine Elms–Vauxhall–
Battersea megaproject 83–84; of space,
in London 18–39; in urban change
18–19
Govindia, R. 87
Graham, S. 8, 153, 154
Grainger PLC 182
Granger, R. 175

Greater London Authority (GLA) 34, 61, 72, 95, 100–101, 151, 195, 220, 222, 234, 239; *Economic Impact of Brexit on London, The* 241; establishment of 35–37; Towards a London Plan 72
Greater London Authority Act of 1999 48
Greater London Council (GLC) 5, 31–32, *31*, 36, 53, 56–57, 63, 219, 220, 241; abolition of 34; establishment of 31
Greater London Development Plan (GLDP) 32, 33, 57, 239, 240; tribulations of 31–33
Greater London Government Act of 1963 31, 32, 36, 53
Great Exhibition of the Arts and Industry (1851) 68, 128, 150
Greenwich Peninsula 60, 64, 73, 74, *75*
Greer, J. 77
Gresham, T. 52
Guildhall School of Music and Drama 70

Haarlem 92
Hackney 37, 38, 49, 60, 95; Calvert Avenue, upgrading of 176; Calvert Street 176–178, *177*; Hackney Wick and Fish Island 139
HafenCity project, Hamburg 66, 145
Hall, P. 13, 33, 61–62, 96, 220, 224, 239; *Cities in Civilization* 62, 77, 92, 143; *Containment of Urban England, The* 29, 240
Hamburg 1, 96, 218; HafenCity project 66; mobilisation of space 224
Hammersmith 32, 58, 60, 88
Hamnett, C. 5, 7, 170, 173, 209, 215, 236; *Unequal City: London in the Global Arena* 70, 170
Hanoi 121
Haringey 61
Harlesden 49
Harris, A. 58, 62, 154, 169, 215
Harrison, B. 7
Harvey, D. 10, 14
Heathrow 33, 59, 228
Hermitage 127
Herzog, J. 81
Heseltine, M. 64, 77, 238; agenda 33–35
higher education 190–191
Highgate 98; gentrification in 178, 180; Highgate Village, London N6 *178*; upgrading in 193
Hill, D. 115
Hirsch, A. 130

Ho, K. C. 10, 15; *Culture and the City in East Asia* 93
Holt, L. 190
Hong Kong 5, 6, 42, 74, 90, 220; gentrification in 193
Hopkins, K.: *Mansion Global* 183–184
House, J. W. 239
Housing of the Working Classes Act, The 206
housing supply/rents 190–191
Hoxton Square 113, 156, *157*; social upgrading 176
HSBC Bank 57
Huguenot refugees, in Spitalfields 21–22
human capital 59, 145, 155, 167, 226, 227
Hunterian Museum 107
Hutton, T. A. 15; *Cities and the Cultural Economy.* 242; *New Economy of the Inner City, The* 242
Hyde Park 68, 122

Ibaraki Prefecture 144
IDE *see* Innovative Design Engineering (IDE) program
imaginaries 53–55, *55*
immigrant communities 198–199
Imperial College of Science, Technology and Medicine 61, 149, *150*, 152, 159, 228; MRC Centre for Global Infectious Disease Analysis 168
Indergaard, M. 123, 229, 237; *Silicon Alley: The Rise and Fall of a New Media District* 11
India 90
industrial development 23–27, *24–27*, **28**
industrial districts, gentrification in 174–178
industrial gentrification 172, 195
industrial urbanism 237
information sharing 239
innovation 53–55; centres, resurgence of cities as 144–146; economy 26; as social construct 161–163; twenty-first-century 146–167; economy, recasting space, place and territory for 145–146; institutional platform for 149–152, *150*; place-based 237; spaces of: processes and exemplars 152–153, *153*; specialisation 153–154; splintering 153–154; succession 153–154; as trajectory 155–159, *157, 158*; in transportation 147–149
Innovative Design Engineering (IDE) program 150

264 Index

Intel 159
internal markers of change, within
London 3–4
Islamabad 6
Isle of Dogs 38, 53, 57, 58, 73, *75*, 136,
219, 238
Islington 35, 37, 95, 151
Islington Borough Council 71

Jacobs, J. M.: *Edge of Empire* 10, 125, 182
Jakarta: mobilisation of space 223
James Keiller & Sons 186
Jarvis, H. 46, 162
Jay, O. 206
Jeffries, S. 101
Jessop, B.: *State-Space* 20
Jewish Museum 105
Johannesburg 8, 14; mobilisation of
space 223
John, P. 115
Johnson, B. 20, 36, 38, 84, 182
Jones, E. 195, 228
Jones, M.: *State-Space* 20
Jones Lloyd 52
Jopling, J. 113

Kaika, M. 53, 231
Ka-Shing, L. 65
Keil, R. 8
Kensington 37, 60, 216; gentrification in
178; upgrading in 193
Kentish Town 38
Khan, S. 36
Kilburn 13, 96, 198
Kim, W. B.: *Culture and the City in
East Asia* 93
Kindertransport 130, 230
King, A. D. 4; *Writing the Global City*
214–215
King George V Dock 74
King's College London 191, 228
King's Cross 61, 68, 90, 143, 149, 230;
Central St Martins Art and Design
College 71, *72*; redevelopment of 70–72
Kolko, J. 123
Kollewe, J. 85–87
Kong, L. 10
Krätke, S. 8, 152; 'New economies, New
spaces' 145
Krippner, G. 47
Kuala Lumpur 14; mobilisation of
space 224
Kynaston, D. 42, 52
Kyoto 143, 145

Labour Borough Council 77
Labour Council 231
Labour Party 5, 219
Lagos 6
Lambeth 151
land use policies 12
Latham, R. 233
Lavender Hill, South Battersea,
Wandsworth 180–181, *181*
LCC *see* London County Council (LCC)
LDDC *see* London Docklands
Development Corporation (LDDC)
Leamouth Peninsula 75
Leather Market 111
Leeds 96
Lees, L. 7, 87, 170, 172, 215, 236
Lefèbvre, H. 20
Leonard Street 157, *158*
Lewisham 151
Lewisham Council 137
Lewis, R. 7, 153
Ley, D. 7, 173; *New Middle Class and the
Remaking of the Central City, The* 170
Leyshon, A. 44
Liverpool 131
Liverpool FC 131
Livingstone, K. 78–79
Lloyd, R. 15, 123, 160, 175–176, 229
LOB *see* Location of Office Bureau (LOB)
Location of Office Bureau (LOB) 30, 31
Logan 44
London: as centre of capital and finance
41–43; change and continuity in
52–53; continuity and contradiction
in development 4–6; gentrification in
171–172; within global urbanisation
circuits 1–2; governance of space in
18–39; growth and change: in the
modern era 22–23; spatial registers of
7–10; half-century of restructuring in
235–236; internal markers of change
within 3–4; in international arena
220–221; positionality 170–171; as site
and exemplar of financialisation 47–50;
as site of change 218–219; upgrading in
172–183; urban megaprojects 68–69;
see also individual entries
London County Council (LCC) 2, 5,
22–23, 31, 53, 149, 195, 206, 219
London Docklands Development
Corporation (LDDC) 35; establishment
of 63–64
London Exposition of 1851 122
London Government Act 1963 185

Index **265**

London Office of Data Analytics 151
London Office of Technology &
  Innovation 151
London Olympic Games: 1908 68, 130;
  1948 49, 68, 130, 134; 2012 49, 58, 59,
  64, 68, 130, 139, 140, 146, 166, 219,
  222, 237
London Plan, The 383
London School of Economics and
  Political Science (LSE) 149, 228
London South Bank University 151
London South East Region 19, 29, 30
London Transport Museum 107
'London – World City 1800–1840'
  exhibition, Essen 4
Long Acre 107
Los Angeles 10, 103; mobilisation of space
  224; postmodern urbanism in 14
Lothbury 52
Lots Road Power Station 82, 83, 243
Louvre 127
Lowry, L. S.: 'Coming Home from the
  Mill,' 119; 'Going to Work' 119
LSE *see* London School of Economics and
  Political Science (LSE)

MacLeod, G.: *State-Space* 20
Manchester 96, 131, 143, 144; as site of
  conviviality 122
Manchester United 131
Manhattan 120
Manners, G. 239
Manthorpe, R. 192
Marina Bay comprehensive
  redevelopment project, Singapore 66
markets 196; changing role of 38–39;
  convivial city 124–127, *126, 127*
Martins, J. 143, 161–163, *161*
Marvin, S. 153, 154
Marx, K. 5, 102, 122, 207
Marx Memorial Library 208
Marylebone: BNP-Paribas financial
  complex *51*
Mason, N. 109
Massey, D. 2, 7, 29, 59; *Regional Division
  of Labour, The* 30; *Spatial Divisions of
  Labour* 241; *World City* 219
Matthews, W. 233
Mayfair 57; landscapes of fund manage-
  ment firms *51*; upgrading in 193
McCann, P. 59
McDowell, L. 14
McGee, T. 226
Mckenzie, M. 15, 102

Meegan, R. 7
megalopolis 11, 226
metropolis: cultural signifiers in 213–217;
  frontiers of upgrading in 180–181;
  geography of 95–97; Millennial
  Metropolis 6–7, 13, 15, 17, 213, 223,
  232, 238–243; money 40–62; rescaling
  territory in 11–15; social resiliency in
  213–217
Metropolitan Board of Works 23, 179
Metropolitan Green Belt 22, 27
Microsoft 144, 159, 168
Middle East 65, 74, 90
Middlesex 33
Middlesex University 134; student
  housing for 184, *185*
Milan 6, 8, 20, 96, 103, 228; mobilisation
  of space 224; social upgrading 175;
  studentification 190
Millennial Metropolis 6–7, 13, 15, 17
Millwall 23, 74
Millwall FC 136–137, *136*
Minato Mirai redevelopment project,
  Yokohama 66
Mission Bay project, San Francisco 66
mobilisation of space, in global city 223–224
modernity 196
Modhin, A. 234
Molotch 44
money metropolis 40–62
Montreal 144
Moore, C. 34, 49
Morris, W. 5; *News from Nowhere* 87
Moscow 13, 220
Mumbai 6, 8, 14, 42, 145, 154, 220;
  gentrification 169–170; mobilisation of
  space 223; zonal structure 59
Muscarà, L 11
Museum of London 70
Museum of Natural History 128
Museum of Science 128

Nathan, M. 143, 163–167
National Gallery 98, 127, 228, 230
National University of Singapore 242
National Westminster Bank Building 53
Natural History Museum 98
neighbourhood festivals 137–141
Nelson, H. 130, 230
neoliberal era, London in 33–35
neoliberalism 4–6, 8, 10, 12, 19, 33–35,
  38, 40, 41, 43, 59, 88, 163, 169, 173,
  214, 215, 217, 220–222, 235, 238,
  241, 242

266 Index

Newcastle United 131
Newham 60, 77
Newham Council 186
New Wembley Stadium, London
    Borough of Brent *135*, 183, 184
New York 2, 3, 6, 8, 13, 35, 42, 98, 103,
    140, 144–146, 154, 166, 167, 169,
    218, 220, 228, 240; financialisation
    47; gentrification 170, 172, 193;
    low-income housing in 192;
    Mid-Manhattan economy 11; as site of
    conviviality 120, 123; social upgrading
    175, 176; studentification 190; urban
    megaprojects 69; zonal structure 59
Nine Elms–Vauxhall–Battersea
    megaproject 12, 16, 60, 64, 73, 79, *81,
    82*, 89, 194, 224, 228, 238, 241, 243;
    multilayered contours of 83–88; agency,
    media and reportage perspectives
    84–88; metanarratives 83–84
North Islington 61
North Woolwich 74

ODP *see* Office Development Permit
    (ODP) system
office development 30–31
Office Development Permit (ODP) system
    30, 31
Office of the High Commissioner for
    Human Rights 182
Olds, K. 65
O'Loughlin, M. 201
Olympia & York 35
online reservations 191–193
Oxford-Cambridge Boat Race 131

Paccoud, A. 172–173
Pacific Rim megaprojects 65
Palazzo Pitti 127
Palmer, J. 239
Paris 1, 4, 6, 10, 13, 14, 17, 20, 41, 92,
    96, 98, 103, 140, 220, 226; everyday
    globalisation in 198, 240; gentrification
    in 193; mobilisation of space 224;
    as site of conviviality 120– 123;
    studentification 190; symbolism of 129
Parker, S. 69
Park Royal 96
Peckham High Street 180, 216;
    cosmopolitan cultures in twenty-first
    century 199–202, *200*; Manze Eel and
    Pie Shop *201*
Peck, J. 29
Pepys, S. 232

Peters, W. 71, 72
Pevsner, N.: *London South* 111
Philip II 21
Piano, R. 64, 76–79, *78*, 231
Pike, A. 164
Pittsburgh: as site of conviviality 122
place 9–10; to London narrative of
    growth and change, centrality of
    13–15; reshaping of 147–149; in urban
    change 10
place-based innovation 237
place-branding 163–167
plague 231–234
planning institutions 27, 29–33
platform effects 227–228
policy problems 33
policy reports 240–242
political economy 12
political mobilization of space 222
political values 43–47
Porter, R. 215
Portobello Saturday Market, Notting Hill
    125, *126*
Port of London: collapse of 1, 5, 29
postcolonialism 238–239
postindustrialism 19, 47, 123, 139,
    195–217; legacy effects of 144–145;
    megaprojects and narratives of 65–66
Pratt, A. 46, 159, 162, 172
Prescott, J. 83
PriceWaterhouse 77, 231
production of space 21–25, 119–121
Project Formation Architects 83
Proll, A.: *Goodbye to London: Radical Art
    and Politics in the 70s* 93
property markets 43–47
Pueblito Paisa Market 182, 183
Punter, J. 65

Quai d'Orsay 127
Queen Anne's Mansions 23
Queen Elizabeth Olympic Park 139
Queen Mary University 159
Queen's Arms 71
Queensway 199, 216; cultural diversity in
    202–205, *204*

Raco, M. 72, 73
Ramsden, M. 209
Randolph, B. 173
Randstadt 13, 220
Reagan, R. 43
regional development 27, 29–33
rental stock 191–193

Index **267**

Renzo Piano Building Workshop (RPBS) 90
reproduction of space 1–17, 59–62, 222–223
respatialisation, and urban megaprojects 69
restructuring 33
Richmond 98, 193
Rio de Janeiro 8, 140
Roman Colosseum 121
Rome 92; mobilisation of space 224; symbolism of 129
Royal Academy, in Burlington Place 98
Royal Albert Dock 74
Royal Albert Hall 99
Royal College of Arts 150, 151
Royal College of Music, South Kensington 99, *99*
Royal College of Surgeons 147
Royal Docks 74, 75
Royal Docks Trust 186
Royal Exchange 52
Royal Opera House 107
Royal Society of Medicine 147
RPBS *see* Renzo Piano Building Workshop (RPBS)
Ruhr 13, 220
Russell Square 107
Rutley, K. F. 114

Sack, R. 11
Saiz, A. 123
Sandercock, L.: *Toward Cosmopolis* 123
San Francisco 3, 8, 15, 35, 98, 107, 145, 154, 155, 162, 164, 166, 216, 217, 220, 240, 242; Mission Bay project 66; urban megaprojects 69
Sao Paulo 14; mobilisation of space 223
Sassen, S. 33, 47, 59, 62, 96, 220; *Global Cities* 13–14
Sawyer, M. 44
Scargill, I. 239
Science Museum 98
SCLSERP *see* Standing Conference on London and South East Regional Planning (SCLSERP)
Scott, A. 225; *Metropolis: From Divisions of Labor to Urban Form* 7
Seattle 66, 103, 145
SEEPC *see* South East Economic Planning Council (SEEPC)
Seoul 6, 20; mobilisation of space 223; zonal structure 59
Seven Sisters Tube Station 182

Shabrina, Z. 192–193
Shaftesbury Theatre 107
Shanghai 3, 8, 14, 42, 55, 59, 103, 166, 218, 220, 228
Shard tower 64, 76–79, *78*, 90, 116, 231
Shaw Theatre 107
Sheppard, F. 32–33
Shepperton 102
Shoreditch 35, 36, 58, 62, 96, 102, 103, 105, 111, 113, 118, 236; cultural innovation 156; football *133*; Goose Island Brewpub, Shoreditch High Street *161*; innovation economy 146, 161; orthogenetic redevelopment of 166; Shoreditch High Street 216; Shoreditch Triangle 156–159, *157, 158,* 176; Silicon Roundabout 159–160, *160,* 164, 216, 219
Shortell, T. 17, 98, 124, 197–198, 202, 204, 240
short-term stays 191
Siem, C. 203
Silicon Alley 123
Silver, S. 186
Silvertown 186
Singapore 6, 42, 55, 90, 145; Chinatown district 15; heritage building restoration in 10; Little India district 15; Marina Bay comprehensive redevelopment project 66; National University of Singapore 242
Sir John Soane's Museum 107
sites and place-based effects 229–230
Smith, D. 190
Smithfield 125
Smith, M. P. 123–124
Smith, N. 169
social capital 1, 2, 59, 97, 123, 137, 139, 140, 142, 168, 224
social class 38, 69–70, 117, 174, 215, 216
social history 196–197
social upgrading 172–179; in the global city, changing fields of 193–194
society: and convivial practices 120
Soho 95, 105
Soja, E. 10, 14, 20, 95
Solnit, R. 15, 217, 229
SOMA *see* South of Market Area (SOMA)
South Bank 60; Festival of Britain in 93
South East Economic Planning Council (SEEPC) 29, 240
South Kensington 95, 96, 98, 111, 143, 149, 152
South Lake Union 144

268 Index

South London 13, 151; gentrification sites in 180–181, 194

South London Innovation Corridor 151

South of Market Area (SOMA) 155, 162, 206, 242

Southwark 23, 33, 37, 60, 79, 102, 151, 205; Bermondsey Antique Market 126, *127*; Borough of Southwark's Planning 78; Southwark Council 76, 115, 191

space/spatial 9–10; in the city, problematising 10; creative, social actors in production of 93–95; of culture 92–93; diversity 196; effects of redevelopment: platform effects 227–228; sites and place-based effects 229–231; territorial effects 228–229; emblematic 169–172; in global city, remaking 221–223; globalisation of 63–65; spatial imaginaries 163–167; of innovation: processes and exemplars 152–153, *153*; specialisation 153–154; splintering 153–154; succession 153–154; mobilisation of 223–234; political mobilization of 222; production of 21–25, 119–121; as register of urban change 224–227; registers of growth and change 7–10; reproduction of 1–17, 59–62, 222–223; transformations of 43–47; capital and 50; vernacular 206–209

Special Procedures of the Human Rights Council 182–183

Spitalfields 21–22, 101, 102; Spitalfields Market 182

Splintering Urbanism 154

stadium development 183–184

Standing Conference on London and South East Regional Planning (SCLSERP) 30, 240

Starbucks 144, 203

Starr, K. 216

state: actors 219–220; changing role of 38–39; neoliberal 218–234; and Nine Elms–Vauxhall–Battersea megaproject 83–84; and urban change 18–19

state-city dialectic 19–21

Stevenson, N. 139–140

Stock Exchange 42, 48

St Pancras 70, 71

St Paul's Cathedral 52

Stratford 60

studentification 190–191

sustainability 137–141

Swyngedouw, E. 88

Sydney: urban megaprojects 69

symbolisms of sites 129–130

Taipei: mobilisation of space 223

Tate Britain 98, 119

Tate & Lyle 186

Tate Modern 60, 64, 84, 90, 98, 113, 114, 222, 231, 238; Tate Modern Gallery 79; Tate Modern Museum 73

Tavistock Institute 152

Tech City project 143, 159–160, 165–166, *165*

technology-based 'new industrial districts' 154–155

territory/territorial 9–10; capitalism 46–47, 62; change, and urban megaprojects 69–70; effects 228–229; financialisation 90, 171, 236; gentrification 171; at the local scale 11–13; to London narrative of growth and change, centrality of 13–15; in metropolis, rescaling 11–15; production of 119–121; reshaping of 147–149; transformations of, capital and 50; of urban change 8–10

Terry Farrell Architects 82

Thames 12, 16, 23, 27, 53, *55*, 60, 64, 74, 83, 84, 87–90, 94, 102, 109, 125, 131, 180, 184, 185, 194, 225, 229, 236, 238; development corridor 76–79; as development corridor and project site 72–73

Thames Gateway plans 12

Thameside megaprojects, multivalent impacts of 89–91

Thamesmead West 74

Thatcher, M. 5, 7, 19, 38, 43, 47, 49, 57, 63, 193, 219, 220, 241; agenda 33–35; policy agenda 45; Thatcherism 35

Theurillat, T. 46–47

Thrift, N.: *Thinking Space* 14

Ticino Bakery 111, 117

Togni, L. 95

Tokyo 6, 13, 42; financialisation 47; mobilisation of space 223

Toronto 35; MaRS research and development cluster 146; as site of conviviality 123

Tottenham 12, 132, 171, 198; Latin Market 181–183

Tottenham Hotspur 131

Tottenham Hotspur Stadium *134*

Tower Bridge 13

Tower Hamlets 60, 77

Tower Hamlets Council 35
Trafalgar Square 95, 127, 130, 230
transnational urbanism 123
transportation 147–149
Travers, T. 85
Trump, D. 20
Tsukuba Science City 144
Turner, S. 142–143
twenty-first-century innovation economy
146–147

Uffizi 127
Unitary Development Plan 2002 53
University College 228
University College London 147,
149, 159
University of Cambridge 226
University of Oxford 61, 226, 232
University of the Arts London, Pimlico
99, 100, 191
University of Toronto's St George
Campus 146
UN Working Group on Business and
Human Rights 182
Upton Park 132, 133
urban change 1–17, 195–197;
appropriation of cultural imaginaries
in 238; comparative study of 240;
diversity in 236–239; governance in
18–19; at the local scale 11–13; place in
10; state and 18–19; trajectories of 8–9
urban-innovation nexus, in history
142–144
urban megaprojects: costs of 67–68; global
diffusion of 66–67; respatialisation and
69; risks of 67–68; territorial change
and 69–70

Vancouver 6, 66, 98, 240; Concord
Pacific project in 65; Victory Square
district 15
van der Zwan, N. 45–46
Vandore, E. 143, 163–167
'Vauxhall Nine Elms Battersea
Opportunity Area Planning
Framework' 383
Veblen, T. 122
Venice: as site of conviviality 121
vernacular cultures 238–239
vernacular space 196–199
Vertovec, S. 124
Victoria and Albert (V&A) Museum 98,
104, 128, 129, 149, 152, 153, 228
Victoria Concordia Crescit 132

Victoria Dock, London Borough of
Newham 62, 64, 73, 74, 83, 90, 171,
229, 243; Britannia Village 75–76,
186–187; Britannia Village project
75–76; capital relayering in 184–189;
high-density residential units 188–189,
189; social upgrading in 184–189
Victorian era (1837–1900) 4, 68, 223
Victorian London 23
Victorian Society 71
Victory Square 102
Vienna 4, 92; convivial practices 120;
as site of conviviality 121
Vijay, A. 87–88; 'Dissipating the
political: Battersea Power Station
and the Temporal Aesthetics of
Development' 87
Voss, G. 143, 163–167

Wachsmuth, D. 192
Wainwright, O. 81
Wainwright, T. 44
Wandsworth 37, 60, 89, 151, 207;
Lavender Hill, South Battersea
180–181, 181
Washington DC 96, 98;
symbolism of 129
Waterloo Station 87, 129
Waterloo Terrace, Barnsbury, London
Borough of Islington 173–174, 174
Weisler, A. 192
Wellcome Collection 107
Wembley 49, 96
west-central London Thameside,
megaprojects within 79–83, 80
West Ham United 131–134, 222
West London 33, 49, 59, 62, 73, 83, 86,
88–89, 96, 131, 224, 234
Westminster 37, 39, 49, 58, 59, 151
Westminster Abbey 94
Westphalian state: problematics of
territory in 11
West Silvertown Community
Foundation 186
Whampoa, H. 82
White Cube Bermondsey Gallery 113,
114, 115
Whitehall 39, 94
White House 101
Wick Festival 139–140
Willesden 49, 96
William I's Tower of London 52
Willmott, P. 12, 13, 137
Wilson, H. 219

Wilson, N. 202
Wimbledon tennis championships 131
Wolverhampton 131
Workspace Group Plc 213
Wren's St Paul's Cathedral 94
Wright, R. 109
Wyly, E. 8

Yeoh, B. 10
Yokohama: Minato Mirai redevelopment project 66
Young, M. 12, 13, 137

Zhong, S. 163
Zola, E. 122
zonal structure: expanded, shaping 59–61, *60*
Zukin, S.: *Cultures of Cities, The* 163; *Loft Living* 163